It's Not Too Late to Learn Computers

An Easy Reference Guide

by

Sean Byerley

author HOUSE™

1663 Liberty Drive, Suite 200
Bloomington, Indiana 47403
(800) 839-8640
www.AuthorHouse.com

First published by AuthorHouse 10/19/04

ISBN: 1-4184-9833-5 (sc)

Printed in the United States of America
Bloomington, Indiana

This book is printed on acid-free paper.

ILLUSTRATIONS BY: RUBEN GERARD

ACKNOWLEDGEMENTS

Thanks to Cindy M., without whom this book might not have happened.

Thanks to George Fernandez for his illustrations.

Thanks to everyone for all your suggestions.

All trademarks and/or copyrights are property of their respective owners.

Buddy List® is a trademark of the America Online® Service and Cooporation and is used with permission.

EZ CD Creator® is a trademark of the Roxio® Coorporation in the United States and/or other countries and is used with permission.

Internet Explorer® is either a registered trademark or a trademark of Microsoft® Corporation in the United States and/or other countries.

It's Never Too Late To Learn Computers is an independent publication and is not affiliated with, nor has it been authorized, sponsored, or otherwise approved by the America Online® Service.

It's Never Too Late To Learn Computers is an independent publication and is not affiliated with, nor has it been authorized, sponsored, or otherwise approved by the eHarmony.com® Coorporation or by Dr. Neil Clark Warren.

It's Never Too Late To Learn Computers is an independent publication and is not affiliated with, nor has it been authorized, sponsored, or otherwise approved by Microsoft® Cooperation.

It's Never Too Late To Learn Computers is an independent publication and is not affiliated with, nor has it been authorized, sponsored, or otherwise approved by Roxio® Coorporation.

MS-DOS® and the MS-DOS® operating system is either a registered trademark or a trademark of Microsoft® Corporation in the United States and/or other countries.

Windows® and the Windows® operating system are either registered trademarks or trademarks of Microsoft® Corporation in the United States and/or other countries.

TABLE OF CONTENTS

ACKNOWLEDGEMENTS vii

CHAPTER 1 – A WORD ABOUT COMPUTERS 1

• WHAT KIND OF QUESTIONS DOES THIS BOOK COVER? 1

• I DON'T NEED A COMPUTER. WHY WOULD I WANT ONE? 1

• BUT, I DON'T TYPE! 2

• WHAT IS SOFTWARE? 2

• ARE ALL COMPUTERS MADE BY THE SAME COMPANY? 2

• DO COMPUTERS TAKE UP A LOT OF ENERGY? 2

• WHAT DO I DO IF I CAN'T AFFORD A COMPUTER? 3

• I RECEIVED SOME MANUALS AND OTHER THINGS WITH MY COMPUT-
ER. I DON'T NEED THEM. CAN I THROW THEM OUT? 3

• I THOUGHT HARDWARE WAS FOUND AT MY LOCAL HARDWARE
STORE. 3

• WHAT DO THESE COLORS ON THE BACK OF MY COMPUTER MEAN? ...3

• WHAT IS SPEECH RECOGNITION SOFTWARE? 4

• OKAY, HOW DO YOU TURN THE THING ON? 4

• WHAT'S THAT BEEPING I KEEP HEARING? 4

• THERE IT GOES AGAIN! 5

• OKAY, NOW THAT I'M FINISHED WITH IT — HOW DO YOU TURN THE
THING OFF? 5

CHAPTER 2 – MORE WHAT IS …? 6

• WHAT IS A DESKTOP COMPUTER? 7

• HEY, I HAVE A QUESTION FOR YOU…HOW DOES THE COMPUTER GET POWER TO RUN? 7

• WHAT IS THE BEST WAY TO LEARN ABOUT HOW TO DEAL WITH COMPUTER PROBLEMS OR INSTALLATIONS YOURSELF? 7

• WHAT IS A TOWER? 8

• WHAT IS A MOTHERBOARD? 8

• WHAT IS A BIOS? 9

• WHAT IS A PROCESSOR? 9

• THERE ARE SO MANY NUMBERS. HOW DO I RECOGNIZE HOW FAST MY PROCESSOR IS? 9

• WHAT DOES FRONT SIDE BUS MEAN? 10

• WHAT IS CACHE? 10

• HOW MUCH IS A BYTE OF INFORMATION? 10

• HOW MUCH IS A MEGABYTE OF INFORMATION? 10

• HOW MUCH INFO IN A GIGABYTE? 11

• I'D REALLY LIKE TO KNOW WHAT A DRIVE IS. 11

• MY COMPUTER HAS A FAN IN IT? 11

• WHAT IS THIS ALPHABET ON MY COMPUTER? 11

• WHAT'S THIS ITEM CALLED A FLOPPY DISK? 12

• I HAVE SEEN SOME DISKS, BUT THEY ARE NOT "FLOPPY" AT ALL. 12

• WHAT IS RAM? 12

• I'M AT A COMPUTER SHOW AND I WANT TO BUY THE BEST / FASTEST MEMORY POSSIBLE? WHAT DO I BUY? 12

• WHAT IS A HARD DRIVE / HARD DISK / FIXED DISK DRIVE? 13

• WHY DO THEY CALL IT A HARD DRIVE / HARD DISK? 13

• THERE IS ANOTHER LITTLE GREEN/RED/YELLOW FLICKERING LIGHT ON MY COMPUTER. WHAT DOES IT MEAN? 13

• WHAT'S THE DIFFERENCE BETWEEN THE HARD DRIVES THAT SAY "5400" AND THOSE THAT SAY "7200"? 13

• IS HAVING YOUR HARD DRIVE "CRASH" A BAD THING? 14

• WHAT IS A MENU? 14

• WHAT'S THE DEFINITION OF A FILE? 14

• I'M NOT GONNA GO OUT TO MY LOCAL OFFICE SUPPLY STORE FOR A FOLDER. 14

• OKAY — NOW THAT I KNOW WHAT A FLOPPY DISK IS, HOW DO I FORMAT IT? 14

• WHAT IS WORD PROCESSING / A WORD PROCESSING PROGRAM? 15

• WHAT IS A CURSOR? 15

• WHAT IS A VIDEO CARD? 15

• TELL ME MORE ABOUT VIDEO CARDS. 15

• WHAT IS A VIDEO CAPTURE CARD? 16

• WHAT IS "TV OUT"? 16

• WHAT IS A PROMPT? 16

• WHAT IS A MONITOR / FLAT SCREEN MONITOR? 16

• AND NOW, GOOD PEOPLE, THE DEFINITION OF A CD-ROM. 17

• I'M TIRED OF WRITING WITH MY PEN ON MY CD-ROM'S. ISN'T THERE A WAY TO MAKE A TYPE-WRITTEN LABEL? 17

• WHAT IS A SPINDLE OF CD-ROMS? 18

• WHAT IS A CD-ROM DRIVE? 18

• WHAT IS A MINI DISC? 18

• WHAT IS A JEWEL CASE?...18

• WILL MY COMPUTER RUN WITHOUT A HARD DRIVE? 18

• WHAT IS A KEYBOARD? 19

• WHAT ARE KEYSTROKES? 19

• WHAT ARE ALL THESE EXTRA KEYS ON MY KEYBOARD? 19

• IS IT POSSIBLE TO USE THE "ENTER" OR "RETURN" KEY TO HIT THE "OK" BUTTON INSTEAD OF TIRING OUT MY HANDS?................20

• WHAT DOES PRESSING "CONTROL ALT DELETE" DO? 22

• THE "NUMERIC KEYPAD" AND THE "NUMBER LOCK" 24

• DOES IT MATTER IF MY "NUMBER LOCK" IS ON OR NOT? 24

• WHAT IS THE DIFFERENCE BETWEEN THE "NUMERIC KEYPAD" AND THE NUMBERS AT THE TOP OF THE KEYBOARD? 24

• WHAT DOES THE "/ " KEY MEAN ON THE "NUMERIC KEYPAD"? 25

• WHAT DOES THE " * " MEAN ON THE "NUMERIC KEYPAD"? 25

• WHAT DOES "+" MEAN ON THE "NUMERIC KEYPAD"? 25

• WHAT DOES THE "-" MEAN ON THE "NUMERIC KEYPAD"? 25

• MY KEYBOARD IS SITTING TOO FLAT. IS THERE ANY WAY TO PROP IT UP A LITTLE BIT?..25

• WHAT IS A MOUSE? 26

• WHAT IS A MOUSE PAD? 26

• WHAT IS A SERIAL PORT?..26

• WHAT DOES THE LEFT MOUSE BUTTON DO? 26

• WHAT DOES THE RIGHT MOUSE BUTTON DO?..........................27

• WHAT IS "RIGHT CLICKING" ON SOMETHING? 27

• WHEN I PUSH THE RIGHT MOUSE BUTTON ON MY MOUSE, LITTLE

MENUS COME UP. WHAT ARE THEY?..27

• WHAT DOES THAT LITTLE WHEEL ON THE TOP OF MY MOUSE DO? 27

• WHAT IS USB?...27

• WHAT IS A USB PORT?...28

• WHAT IS A USB KEY?...28

• WHAT IS A USB HUB? 28

• WHAT IS A MODEM? 28

• WHAT IS KPBS?...29

• MY MODEM HAS TWO PHONE JACKS IN THE BACK OF IT. WHY?..........29

• IF IT'S A 56K MODEM, HOW COME IT NEVER CONNECTS AT 56K? 29

• WHAT IS THE DIFFERENCE BETWEEN A REGULAR MODEM CONNEC-
TION AND A CABLE MODEM CONNECTION? 29

• WHAT IS A DVD?..29

• WHAT IS A DVD-ROM? 30

• IS A CD/CD-ROM/DVD/DVD-ROM BREAKABLE OR SCRATCHABLE?.......30

• TELL ME A LITTLE ABOUT DVD-ROM DRIVES...............................30

• CAN I PLAY DVD MOVIES ON MY COMPUTER? 30

• WHAT IS A SCANNER?..30

• WHAT IS A PRINTER? 31

• WHY DO I NEED TWO CARTRIDGES FOR MY PRINTER (DUAL CAR-
TRIDGE PRINTERS)? 31

• WHAT IS A RADIO BUTTON? 31

• MY INK CARTRIDGE IS LOW ON INK. IS THERE A WAY TO CONSERVE
INK? 31

• I PRINTED SOMETHING THAT WAS CUT OFF AT THE RIGHT SIDE OF

THE PAGE. 32

• IN A PREVIOUS QUESTION, YOU MENTIONED SOMETHING ABOUT A DIFFERENT KIND OF "TAB" THAN THE ONE ON THE KEYBOARD. WHAT IS THIS KIND OF "TAB"? 32

• I DECIDED THAT I DIDN'T WANT TO PRINT SOMETHING AND IT'S ABOUT TO PRINT. 33

• WHAT IS A PAPER JAM AND WHY DOES IT ALWAYS SEEM TO HAPPEN AT TO ME AT THE WRONG TIME? 33

CHAPTER 3 – NAVIGATING THE MICROSOFT® WINDOWS® XP OPERATING SYSTEM...34

• WHAT DOES INSTALL MEAN? 34

• WHAT IS THE MICROSOFT® WINDOWS® OPERATING SYSTEM DESK-TOP? 34

• WHAT IS AN ICON? 34

• HOW DO I BEGIN USING THE MICROSOFT® WINDOWS® XP OPERATING SYSTEM? 34

• WHAT IS THE MICROSOFT® WINDOWS® XP OPERATING SYSTEM TASK-BAR? 35

• I DON'T WANT THE MICROSOFT® WINDOWS® OPERATING SYSTEM TASKBAR AT THE BOTTOM OF MY SCREEN. WHAT DO I DO TO TEMPO-RARILY GET RID OF IT? 35

• WHAT IS THAT HOURGLASS FOR? 35

• WHAT IS THAT LITTLE ARROW IN THE LOWER RIGHT HAND COR-NER? 35

• WHY DO THEY CALL IT THE MICROSOFT® WINDOWS® OPERATING SYSTEM AND DO I HAVE TO CLOSE ONE MICROSOFT® WINDOWS®-BASED APPLICATION BEFORE I CAN OPEN ANOTHER?.............36

• WHAT IS A DROPDOWN MENU? 36

• I WANT TO SAVE OR VIEW A DOCUMENT ON ANOTHER DRIVE. HOW DO I CHANGE THE DRIVE OR FOLDER I AM LOOKING AT? 37

• I NEED TO INSTALL A NEW COMPUTER PROGRAM ONTO MY COMPUTER. WHAT DO I DO? 37

• THE MICROSOFT® WINDOWS®-BASED APPLICATIONS LISTED ON MY COMPUTER ARE OUT OF ORDER. IS THERE ANY WAY TO ALPHABETIZE THEM? 38

• WHAT IS THE *RECYCLE BIN*? 38

• HOW DO I EMPTY THE *RECYCLE BIN*? 38

• I DELETED SOME FILES FROM MY COMPUTER, BUT THE SPACE ON MY HARD DRIVE IS STILL TAKEN UP? WHAT'S WRONG? 39

• WHAT DOES "DEFAULT DRIVE, FOLDER, OR BUTTON" MEAN? 39

• WHAT ARE THOSE THINGS IN THE BLUE BAR IN THE TOP RIGHT HAND CORNER OF MY SCREEN? 39

• WHEN I STOP MY MOUSE ARROW OVER SOMETHING, IT SHOWS LITTLE BOXES WITH WORDS IN THEM. WHAT DOES THIS MEAN? 40

• I'M IN THE MICROSOFT® WINDOWS® XP OPERATING SYSTEM AND I'VE INSTALLED A MICROSOFT® WINDOWS®-BASED APPLICATION ON MY COMPUTER. NOW HOW DO I USE IT? 40

• I WANT TO FIND A FILE ON THE COMPUTER. HOW DO I DO THAT? 41

• WHAT IS "MY COMPUTER"? 41

• WHAT IS "MY DOCUMENTS"? 42

• WHAT IS THE *"MICROSOFT® WINDOWS® EXPLORER®"* WINDOWS®-BASED APPLICATION? 42

• I DOUBLE CLICKED ON A FOLDER TO SEE WHAT WAS INSIDE IT. I WENT INTO THE WRONG FOLDER. WHAT CAN I DO? 43

• HOW CAN I CHECK TO SEE HOW MUCH SPACE IS TAKEN UP ON MY HARD DRIVE? 43

• I'M IN *THE "MICROSOFT® WINDOWS® EXPLORER®"* WINDOWS®-BASED APPLICATION OR *"MY DOCUMENTS"*. I RIGHT CLICK ON A FILE AND ONE OF THE MENU OPTIONS SAYS, "OPEN WITH". WHAT DOES "OPEN WITH"... DO? 43

• WHAT DOES DRAGGING AND DROPPING A FILE MEAN? 44

• HOW DO I DRAG AND DROP A FILE? 44

• I'M TRYING TO COPY FROM ONE FOLDER TO ANOTHER IN THE *"MICROSOFT® WINDOWS® EXPLORER®"* WINDOWS®-BASED APPLICATION. IS THERE A WAY I CAN SEE BOTH FOLDERS AT THE SAME TIME?45

• HOW DO I COPY AND PASTE OR CUT AND PASTE A FILE? 46

• WHAT IS SCREEN RESOLUTION? 46

• I NEED TO CHANGE MY SCREEN RESOLUTION. 46

• WHAT DOES REFRESH RATE MEAN AND HOW DO I CHANGE IT? 47

• I WANT TO RENAME A FILE I HAVE ALREADY SAVED. 47

• WHAT IS THE "CONTROL PANEL"? 48

• WHAT IS A SCREEN SAVER? 50

• HOW DO I CHANGE MY SCREEN SAVER? 51

• WHAT IS A FONT? 52

• WHAT IS THE "DEVICE MANAGER" AND HOW DO I GET TO IT? 57

• HOW DO YOU CLOSE THE MICROSOFT® WINDOWS® XP OPERATING SYSTEM? 58

• WHAT HAPPENS IF I DON'T SHUT DOWN MY COMPUTER PROPERLY? 58

• OKAY, I ACCIDENTALLY SHUT DOWN MY COMPUTER WRONG. 59

• WHAT DOES THE "LOG OFF" BUTTON DO? 59

• WHAT DOES THE "STAND BY" BUTTON DO? 59

• HOW DO I BRING MY COMPUTER OUT OF "SLEEP MODE"? 59

• WHAT DOES THE "RESTART" BUTTON DO? 60

• DO I NEED TO HAVE A SEPARATE ORIGINAL COPY OF THE MICROSOFT® WINDOWS® XP OPERATING SYSTEM FOR EACH COMPUTER I OWN? 60

CHAPTER 4 – THE ENTERTAINMENT **61**

• WHAT DO I DO IF I DON'T KNOW WHAT MICROSOFT® WINDOWS®-BASED APPLICATION TO OPEN A PARTICULAR FILE WITH? 61

• WHAT IS THE MICROSOFT® MICROSOFT® WINDOWS® XP OPERATING SYSTEM MEDIA® PLAYER? 61

• I WANT TO OPEN A PARTICULAR FILE IN THE MICROSOFT® WINDOWS® XP OPERATING SYSTEM MEDIA® PLAYER, AND I NEED TO FIND THE FILE MENU. WHERE IS IT? 61

CHAPTER 5 — BUYING A NEW COMPUTER **63**

• UPGRADABILITY? TELL ME MORE. 63

• HOW DO I KNOW WHAT TO GET ON MY NEW COMPUTER? 63

• HOW DO I HOOK UP MY SOUND CARD TO MY CD-ROM DRIVE SO I CAN LISTEN TO MUSIC CD'S? 66

• I DON'T HAVE ANY VOLUME CONTROLS ON MY SPEAKERS. IS THERE A WAY TO TURN DOWN THE SOUND? 66

• I WANT TO TURN THE SOUND OFF COMPLETELY (TEMPORARILY, OF COURSE) 67

• I TRIED TO INSTALL A NEW PROGRAM ON MY COMPUTER. THE COMPUTER IS WORKING FINE, BUT THE NEW PROGRAM WON'T WORK? 67

• WHAT IS A SURGE PROTECTOR? 68

• THE MICROSOFT® WINDOWS® XP OPERATING SYSTEM ASKED ME WHAT PROGRAM I WANTED TO OPEN A PICTURE FILE WITH. I CHOSE ONE FROM THE LIST. WHEN I DOUBLE CLICKED ON THE PICTURE TO OPEN IT, I SAW NOTHING. CAN I CHOOSE ANOTHER MICROSOFT® WINDOWS®-BASED APPLICATION 68

CHAPTER 6 – THE INTERNET 70

• WHAT IS THE INTERNET? 70

• I HAVE A COMPUTER AND IT IS TURNED ON. DOES THIS MEAN I AM CONNECTED TO THE INTERNET? 70

• WHAT DOES "WWW" MEAN? 70

• WHAT DOES ".NET" MEAN? 70

• WHAT DOES ".ORG" MEAN? 70

• WHAT IS E-MAIL? 71

• WHAT IS A WEBSITE/WEBPAGE? 71

• WHAT IS A BROWSER? 71

• WHAT IS A SCREEN NAME? 71

• WHY DO I WANT AN INTERNET MEMBERSHIP? 71

• IS THE INTERNET SAFE FOR ME AND/OR MY CHILDREN? 71

• WHAT DOES IT MEAN WHEN SOMEONE TELLS YOU TO LOG ON TO THE INTERNET? 72

• HOW DO I LOG ON TO THE INTERNET? 72

• WHY DO I NEED A PASSWORD? 72

• WHAT SHOULD I DO IF I HAVE FORGOTTEN MY PASSWORD? 73

• WHY DO I NEVER SEEM TO CONNECT TO THE INTERNET AT THE SAME SPEED? 73

• OK, I'M CONNECTED. WHERE FROM HERE? 74

• CAN I TALK ON THE PHONE AND BE LOGGED ONTO THE INTERNET AT THE SAME TIME? 75

• WHAT DOES DOWNLOAD MEAN? 75

• WHAT DOES UPLOAD MEAN? 75

• HOW CAN I TELL HOW MUCH LONGER WILL THIS PAGE TAKE TO LOAD? 75

• WHAT DOES TRANSFER RATE MEAN? 75

• WHAT INTERNET PROVIDER SHOULD I USE? 75

• WHAT IS A VIRUS? 76

• WHAT ARE "TEMPORARY INTERNET FILES"? 76

• HOW DO I CLEAR MY INTERNET HISTORY? 76

• WHAT IS A FIREWALL? 77

• HOW DO I GET AN E-MAIL ADDRESS? 77

• HOW DO I SEND AN E-MAIL? 77

• WHAT IS AN ATTACHMENT? 78

• I WANT TO SEND A NEW PICTURE OF THE GRANDKIDS TO MY FAMILY. HOW DO I ADD THIS TO MY E-MAIL? 78

• I OPENED AND READ AN E-MAIL AND NOW I CANNOT FIND IT AGAIN...78

• I DOWNLOADED A PICTURE LAST NIGHT AND I WANT TO LOOK AT IT AGAIN. DO I HAVE TO DOWNLOAD IT AGAIN TO LOOK AT IT. 79

• WHAT IS A "FILING CABINET"? 79

• WHAT IS A LINK? 80

• WHAT IS A SEARCH ENGINE? 80

• I'VE HEARD I CAN KEEP A LIST ON MY COMPUTER OF THE WEBSITES I LIKE BEST. HOW DO I ADD SOMETHING TO MY FAVORITE PLACES LIST? 81

• HOW DO I REMOVE SOMETHING FROM MY FAVORITE PLACES LIST? .81

• WHAT IS RELOADING? 81

• WHAT ARE COOKIES? 81

• I NEED TO RE-WORK MY SETTINGS FOR ACCEPTING COOKIES. HOW DO I DO THIS? 82

• I ADJUSTED MY SETTINGS TO ACCEPT COOKIES, AND I'M STILL HAVING TROUBLE. WHAT ELSE CAN I DO? 82

• CAN I BUILD MY OWN WEBSITE? 83

• WHAT IS A SERVER? 83

• WHAT DO I NEED TO BUILD MY OWN WEBSITE? 83

• WHAT DOES THAT LONG INTERNET ADDRESS MEAN? 83

• WHAT IS AN INTERNET CAMERA? 84

• OKAY, I'M DONE. NOW HOW DO I LOG OFF? 85

• IS THERE ANOTHER WAY TO LOG OFF THE INTERNET? 85

• I HAVE A GAME THAT CAN ONLY BE PLAYED OVER THE INTERNET. 85

• THAT'S NICE. CAN I PLAY THIS GAME ON MY OWN COMPUTER WHEN I'M LOGGED OFF OF THE INTERNET? 85

• WHAT IS AN INSTANT MESSAGE? 85

• HOW DO I SEND AN INSTANT MESSAGE? 86

• CAN I SEND INSTANT MESSAGES BACK AND FORTH WITH MORE THAN ONE PERSON AT THE SAME TIME? 86

• IS THERE A WAY TO TELL IF A PARTICULAR SOMEONE IS ONLINE SO I CAN SEND THEM AN INSTANT MESSAGE? 86

• I WANT TO USE THE AMERICA ONLINE® SERVICE BUDDY LIST® FEATURE, BUT I DON'T HAVE ANY PEOPLE ADDED YET. HOW DO I ADD PEOPLE TO THIS LIST? 87

• HOW DO I DELETE SOMEONE FROM THE AMERICA ONLINE® SERVICE BUDDY LIST® FEATURE? 87

• OKAY. I'M USING THE AMERICA ONLINE® SERVICE BUDDY LIST® FEATURE AND I SEE THAT THEY ARE ONLINE, HOW DO I SEND HIM OR HER AN INSTANT MESSAGE? 88

• I'M DOING FINANCIAL TRANSACTIONS ONLINE. IT SAYS I NEED A BANK ACCOUNT. I DON'T WANT TO GIVE OUT THAT INFORMATION ONLINE. ...88

CHAPTER 7 – WORKING WITH THE MICROSOFT® WORD® WINDOWS®-BASED APPLICATION **89**

• WHAT IS THE MICROSOFT® WORD® WINDOWS®-BASED APPLICATION? 89

• WHAT IS HIGHLIGHTING? 89

• HOW DO I START A NEW DOCUMENT? 89

• HOW DO I OPEN A FILE? 90

• HOW DO I SAVE A FILE? 90

• I SEE THAT THERE IS TWO CHOICES TO SAVE A DOCUMENT – "SAVE" OR "SAVE AS". WHAT'S THE DIFFERENCE? 91

• HOW DO I PRINT A DOCUMENT? 91

• I JUST TYPED A LETTER AND I WANT TO RUN A "SPELLING AND GRAMMAR CHECK". HOW DO I DO THAT? 92

• I SEE RED AND GREEN LINES IN MY DOCUMENT. 92

• IF I PRINT OUT MY DOCUMENT AS IT IS, WILL THESE RED AND GREEN LINES SHOW UP ON MY PRINTOUT? 92

• I WANT TO USE A DIFFERENT FONT. HOW DO I CHANGE IT? 92

• I WANT TO CHANGE THE FONT SIZE (HOW BIG IT IS). HOW DO I DO THAT? 92

• I WANT TO HAVE MY TEXT IN BOLDFACE. HOW DO I DO THAT? 93

• I WANT TO HAVE MY TEXT IN ITALICS. HOW DO I DO THAT? 93

• I WANT TO UNDERLINE MY TEXT OR UNDERLINE SOME TEXT THAT I DIDN'T UNDERLINE BEFORE. HOW DO I DO THAT? 93

• I WANT TO CHANGE THE COLOR OF MY TEXT. HOW DO I DO THAT?....94

• WHAT IS SCROLLING AND HOW DO I DO IT? 94

• I HAVE A VERY LARGE DOCUMENT THAT I AM EDITING AND I SCROLLED DOWN REALLY FAR, THEN I PRESSED ONE OF MY ARROW KEYS AND THE CURSOR WENT RIGHT BACK TO WHERE IT WAS BEFORE..............95

• I HAD MY FONT SIZE (SIZE OF TEXT) SET TO A LARGER SIZE AND NOW ITS SMALL AGAIN. 95

• THERE IS A PARTICULAR WORD OR PHRASE I WANT TO FIND OR GO TO IN MY DOCUMENT. WHAT IF I WANT TO GO TO A SPECIFIC PAGE? HOW DO I DO THAT? 97

• THE MENUS HAVE BEEN CUT OFF!...97

• WHAT ARE ALL THOSE BUTTONS AT THE TOP? 97

• WHEN YOU ARE SAVING A DOCUMENT AND IT SAYS "SAVE IN THIS OR THAT FORMAT", WHAT DOES IT MEAN? 98

• I HAVE TWO DOCUMENTS I WANT TO MAKE INTO ONE. 99

• IS THERE ANY WAY TO AVOID LOSING A DOCUMENT DUE TO A POWER OUTAGE? 99

• I HAVE TWO DIFFERENT VERSIONS OF THE FILE I'M WORKING ON. HOW DO I TELL WHICH ONE IS THE ONE I WORKED ON LAST?..............99

• WHAT IS A TABLE? 100

• HOW DO I CREATE A TABLE? 100

• WHAT DOES "PRINT PREVIEW" DO? 101

• WHERE IS "PRINT PREVIEW?" 101

• I'M TYPING MY RESUME AND I WANT TO INCLUDE MY PICTURE. CAN I DO THAT IN THE MICROSOFT® WORD® WINDOWS®-BASED APPLICATION? 101

CHAPTER 8 – MAKING FILES SMALLER **103**

• WHAT IS A COMPRESSED FILE (OR FOLDER)? 103

• .... DO I NEED A SEPARATE MICROSOFT® WINDOWS®-BASED APPLICATION OR CAN I MAKE A COMPRESSED FILE (OR FOLDER) IN THE MICROSOFT® WINDOWS® XP OPERATING SYSTEM? 103

• HOW DO I CREATE A COMPRESSED FOLDER FULL OF FILES? (I DON'T HAVE A FOLDER FULL OF FILES YET). 103

• WHAT IF I ALREADY HAVE A FOLDER OF FILES THAT I WANT TO COMPRESS? 104

• HOW DO I DECOMPRESS A COMPRESSED FOLDER? 104

• THERE'S A FILE THAT I FORGOT TO ADD TO THAT COMPRESSED FOLDER! 105

• IS THERE A QUICKER WAY TO COMPRESS A SINGLE FILE TO A NEW COMPRESSED FOLDER? 105

CHAPTER 9 – BURNING CD'S OR DVD'S **106**

• WHAT IS A CD / DVD BURNER AND WHAT DOES "BURNING" A CD OR A DVD, MEAN? 106

• WHAT IS WRITING TO A DISC / DISK? 106

• WHAT IS CD-R? 107

• WHAT IS CD-RW? 107

• WHAT IS DVD-R? 107

• WHAT IS A DVD+R? 108

• WHAT IS A DVD-RW? 108

• CAN I BURN CD'S USING THE MICROSOFT® WINDOWS® XP OPERAT-
ING SYSTEM? 108

• HOW DO I BURN FILES TO A CD OR DVD IN THE MICROSOFT® WIN-
DOWS® XP OPERATING SYSTEM? 108

• WHAT IS FINALIZING A DISK? 109

• WAIT! THERE'S A FILE ON THERE I DON'T WANT WRITTEN TO THE
CD. 109

CHAPTER 10 – DIGITAL CAMERAS AND WORKING WITH PICTURES 110

• WHAT IS A DIGITAL CAMERA? 110

• HOW DO I GET THE PICTURES FROM MY DIGITAL CAMERA TO MY COM-
PUTER? 110

• I CAN MAKE MOVIES WITH MY DIGITAL CAMERA? 111

• WHICH DO YOU RECOMMEND FOR MY DIGITAL CAMERA – DIGITAL
ZOOM OR OPTICAL ZOOM? 111

• WHAT IS A MEMORY CARD? 111

• I NEED TO BUY A MEMORY CARD FOR MY DIGITAL CAMERA. HOW DO I
DETERMINE WHAT SIZE TO BUY? 112

• WHAT'S THE PURPOSE OF A MEMORY CARD READER? 112

• DO ALL DIGITAL CAMERAS USE THE SAME TYPE OF MEMORY
CARD? 112

• WHO AM I GOING TO GET TO TAKE PICTURES OF ME? 112

• WHAT OTHER ACCESSORIES CAN I BUY THAT WILL BE USEFUL? 112

• I NEED TO BUY BATTERIES FOR MY DIGITAL CAMERA. WHERE CAN I
BUY THEM? 113

- I HAVE SOME OLD PHOTOS THAT ARE FALLING APART. CAN COMPUTERS HELP ME TO PRESERVE THEM? 113

- WHAT ARE THUMBNAILS? 113

CHAPTER 11 — TROUBLESHOOTING TOPICS 114

- I TURNED BY COMPUTER ON, BUT I DON'T SEE THE MICROSOFT® WINDOWS® OPERATING SYSTEM... 114

- MY COMPUTER IS FROZEN AND I CAN'T TURN IT OFF! WHAT DO I DO NOW? 114

- I WANT TO SEE WHAT'S INSIDE A FOLDER. I CLICKED ON IT, BUT IT DIDN'T OPEN. WHAT'S WRONG? 115

- I THINK I HAVE A COMPUTER VIRUS. WHAT DO I DO? 115

- MY COMPUTER IS RUNNING FINE, BUT THERE'S NOTHING ON MY COMPUTER SCREEN.. 115

- THE PICTURE ON MY COMPUTER MONITOR IS OFF TO ONE SIDE. 115

- I MOVED MY CURSOR AND I CLICKED IT IN THE LITTLE BOX TO BE ABLE TO TYPE SOMETHING. THE CURSOR IS BLINKING, BUT I PRESS THE KEYBOARD KEYS AND THE LETTERS/NUMBERS DON'T SHOW UP ON THE MONITOR.. 116

- HOW DO I MAKE A BOOT DISK/START-UP DISK (WITHIN THE MICROSOFT® WINDOWS® XP OPERATING SYSTEM)? 116

- I JUST USED MY BOOT DISK TO BOOT MY COMPUTER AND IT BOOTED UP TO A MICROSOFT® MS-DOS® OPERATING SYSTEM COMMAND PROMPT. HOW DO I CHANGE TO DIFFERENT DRIVES ON MY COMPUTER? 117

- MY COMPUTER HAS REALLY SLOWED DOWN. WHAT HAPPENED? 117

- HOW DO I DEFRAGMENT MY HARD DRIVE? 118

- WHAT DO I DO IF MY COMPUTER IS STILL RUNNING SLOWLY AFTER I

HAVE DEFRAGMENTED THE HARD DRIVE? 119

• MY KEYBOARD IS DUSTY (INSIDE AND OUT)! DO YOU HAVE ANY SUG-GESTIONS TO CLEAN IT? 119

• HOW DO I CLEAN MY CD-ROM (THE CD-ROM DISK)? 119

• WHAT'S THE BEST WAY TO KEEP DUST FROM GETTING INSIDE MY COMPUTER? 119

• I CAN'T TURN MY COMPUTER ON AND I NEED TO GET THE CD-ROM OUT OF THE DRIVE. 119

• I JUST TRIED TO INSTALL A NEW VIDEO CARD INTO MY COMPUTER AND NOW MY COMPUTER DOESN'T WORK. 120

• I WENT TO HOOK UP MY MOUSE AND THE CONNECTOR ON THE MOUSE AND THE CONNECTOR ON THE BACK OF THE COMPUTER DON'T MATCH! 120

• I JUST TRIED TO READ THE INFORMATION ON MY FLOPPY DISK AND THE COMPUTER WON'T READ IT! 120

• I JUST BOUGHT A NEW SCANNER AND I CAN'T GET IT TO POWER UP. IS IT BROKEN? 121

• HOW COME MY SCANNER DOESN'T SCAN IN COLOR? 121

• I TRIED TO PRINT SOMETHING AND NOTHING HAPPENED. 121

CHAPTER 12 – THE MICROSOFT® MS-DOS® OPERATING SYSTEM AND THE MICROSOFT® MS-DOS® COMMAND PROMPT 123

• WHAT IS THE MICROSOFT® MS-DOS® OPERATING SYSTEM? 123

• WHAT IS A MICROSOFT® MS-DOS® OPERATING SYSTEM COMMAND PROMPT? 123

• HOW DO I GET TO A MICROSOFT® MS-DOS® OPERATING SYSTEM PROMPT FROM WITHIN THE MICROSOFT® WINDOWS® XP OPERATING SYSTEM? 123

• TELL ME ABOUT SOME OF THE DIFFERENT VERSIONS OF THE MICROSOFT® WINDOWS® OPERATING SYSTEM? 124

CHAPTER 13 – TECHNICAL STUFF **125**

• I NEED A SCREWDRIVER TO USE FOR WORKING ON MY COMPUTER SYSTEM. 125

• I WENT TO A COMPUTER SHOW AND BOUGHT A COMPUTER CASE (TOWER). THERE IS A LOCK ON THE FRONT OF IT. WHAT IS THAT?......125

• I'M GOING TO WORK INSIDE MY COMPUTER. IS THERE ANYTHING I NEED TO DO FIRST? 125

• WHAT IS A PARTITION / PARTITIONING YOUR HARD DRIVE?................125

• WHAT HAPPENS TO THE OTHER HARD DRIVE LETTERS WHEN YOU USE SEVERAL TO PARTITION A HARD DRIVE? 126

• WHEN I PARTITION MY HARD DRIVE, WILL I LOSE THE INFORMATION THAT IS ALREADY ON IT? 127

• OKAY, YOU SAID THERE WAS SOFTWARE FOR PARTITIONING A HARD DRIVE. WHERE DO I GET THE SOFTWARE FROM? 127

• IS THERE ANYTHING ELSE I'M GOING TO NEED WHEN PARTITIONING (OR RE-PARTITIONING) MY HARD DRIVE? 127

• I JUST TRIED TO INSTALL MY NEW CD-ROM DRIVE IN MY COMPUTER, BUT IT STILL ISN'T BEING RECOGNIZED? 127

• HOW MANY PHYSICAL DRIVES CAN I HAVE INSIDE MY COMPUTER?.128

• WHAT IS A BOOT SECTOR? 128

• I HEARD THAT MONITORS PUT OUT A VERY STRONG MAGNETIC FIELD. IS THIS TRUE? 128

AFTERWORD **129**

CHAPTER 1 – A WORD ABOUT COMPUTERS

I know that there are many people out there that want nothing to do with computers. Despite what most people who want to have nothing to do with computers will tell you, your computer will not explode or stop working just because you push the wrong key or something.

You may be saying, "I'm too old to learn computers!" Not so. Whether you're in your 20's, 30's, 40's, 50's, 60's, or 70's, you aren't too old to learn computers.

You might be speculating about how this book came to be. Well, about 1990 or so, people started telling me, "You should write a book!" because I was always (or almost always) able to answer their computer questions. So, here I am seven computers and many friends later, finally putting it down on paper.

Without further ado, here's our first question of many....

• WHAT KIND OF QUESTIONS DOES THIS BOOK COVER?

This book covers questions about PC's, the internet, and many other computer related questions.

• I DON'T NEED A COMPUTER. WHY WOULD I WANT ONE?

Computers are around to make our life easier. Do you write many letters? Good. Let's say, however, that you don't print or handwrite very well. A computer can help you type an extremely neat letter to your sister, favorite aunt or uncle, mother, father, etc., faster and easier than you might have thought possible. Just print it out and send. In addition, I know that many people buy a computer just to send e-mails (see the question, "WHAT IS E-MAIL?") to their relatives.

You can make greeting cards. You can also make an album of all your favorite photos, play movies, music, games, all on your computer.

BASIC WHAT'S WHAT

• BUT, I DON'T TYPE!

You don't have to be able to type to use a computer, although it is helpful.

• WHAT IS SOFTWARE?

Software is the programs (games, word processing programs, etc.) that you put into the computer from the outside of it.

• ARE ALL COMPUTERS MADE BY THE SAME COMPANY?

No. There are many companies out there that make computers.

• DO COMPUTERS TAKE UP A LOT OF ENERGY?

No. Years ago, when home computers were in their infancy, they were a lot louder, a lot larger, and took up a lot more resources, including energy. Today, they take up only a portion of that energy.

• WHAT DO I DO IF I CAN'T AFFORD A COMPUTER?

No problem! You can buy a computer with your credit card and pay it off little by little until you have paid it off completely — and use it while you're doing it.

• I RECEIVED SOME MANUALS AND OTHER THINGS WITH MY COMPUTER. I DON'T NEED THEM. CAN I THROW THEM OUT?

No. You must keep them. The manuals may come in handy later, and the rest are software disks (that you already paid for and might need to use or install the software again). A couple of the disks contain most of the software that came on your computer – including the software that makes it run. So, you must hang on to these items.

• I THOUGHT HARDWARE WAS FOUND AT MY LOCAL HARDWARE STORE.

Well…this is a different kind of hardware. In this case, hardware is everything that is not software (all the internal parts of your computer).

• WHAT DO THESE COLORS ON THE BACK OF MY COMPUTER MEAN?

When you're hooking up your computer, the cables and the hardware should match. That way you don't have to wonder what connects to what.

• WHAT IS SPEECH RECOGNITION SOFTWARE?

Let's say you want to type a letter, but your hands hurt, or you just plain don't feel like it. This software will let you say what you want to say in your letter, and it will end up on your computer screen – typed for you. All you need is a good microphone. Here comes the question I most often come across...

• OKAY, HOW DO YOU TURN THE THING ON?

There is a button on the front of the computer tower. Power buttons for computers have moved around and changed a little bit over the years. On the smaller computers of the early 1980's, the power buttons (or switches as they were then) were mostly small black switches on the back left hand side of the computer. In the late 80's, they were long switches on the front right hand side of the computer. The power switches on today's computers are usually located on the front right hand side of the computer.

• WHAT'S THAT BEEPING I KEEP HEARING?

That beeping is when your computer is booting up. No, it doesn't mean kicking it when it doesn't do what you want it to. Booting up your computer (or your hard drive) means turning it on. When you hear that beep shortly after you turn your computer on... that's the point at which you are booting up your computer.

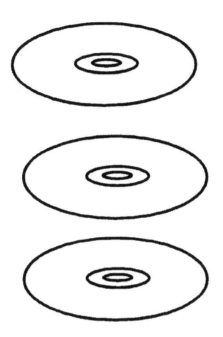

• THERE IT GOES AGAIN!

Making your computer go through the same process again (as when you first turn it on) is called "rebooting" your computer. Sort of like making it turn on again, even though it's already on.

• OKAY, NOW THAT I'M FINISHED WITH IT — HOW DO YOU TURN THE THING OFF?

1. Press the "Start" button
2. Press "Turn Off Computer"
3 Press "Turn Off"

See the question, "HOW DO YOU CLOSE THE MICROSOFT® WINDOWS® XP OPERATING SYSTEM?" You will find it later on in this book.

CHAPTER 2 – MORE WHAT IS …?

Most people new to computers tell me, "I barely know how to turn it on. I don't understand computers." Well, let's start there. A computer is a lot like an automobile. There's a lot going on under the hood before you even start using it.

• WHAT IS A DESKTOP COMPUTER?

A desktop computer is stationary and is meant to be put in one place and kept there — usually at your desk – hence the name.

• HEY, I HAVE A QUESTION FOR YOU…HOW DOES THE COMPUTER GET POWER TO RUN?

Everyone knows you need electricity to make a computer work. What most people don't know or may not know is that you need what is called a power supply. This rectangular object near the top rear of the inside of your tower (your computer case) and supplies all these parts with the power they need to run quickly and properly. Most computers today have power supplies that are at least 250-300 watts.

• WHAT IS THE BEST WAY TO LEARN ABOUT HOW TO DEAL WITH COMPUTER PROBLEMS OR INSTALLATIONS YOURSELF?

For me, the best way was to fiddle. In 1993, I got my first clone (a computer without someone's brand name on it), and I knew that I had to learn more about all this new stuff. I had, for example, purchased a new CD-ROM drive. I wanted to install it myself, so I could get started using it quickly. I opened my computer and looked around to see if everything was there for me to hook it up. I took the CD-ROM drive out of the box, screwed it into place inside the computer so it would not fall out or move, and I looked at the other drives to see how they were hooked up and hooked that one up the same way. Eventually, you get a feel for what you should and shouldn't be doing by yourself. If you think that you shouldn't do this because you may not know how, don't do it – take it to a technician or a computer store, so they can fix your computer or install something new for you.

• WHAT IS A TOWER?

No, not that kind of tower. A computer tower is that thing sitting on the floor (No, not the computer screen, the other tall heavy object) — the computer case. Above you see a diagram of a computer tower.

From top to bottom you see: two CD-ROM or DVD-ROM drives.

Below that you see: a floppy drive.

Below that is: The power button.

To the right of the floppy drive is where you might find a USB port.

Please note that there are many different variations of computer towers.

• WHAT IS A MOTHERBOARD?

Every computer has a motherboard. The motherboard is the "skeleton" of the computer. A motherboard is a flat board inside the computer with a bunch of circuitry on it that is generally secured to the side or bottom of the case with screws, etc. to stay securely in place. It holds many of the other pieces on the inside of the computer together. The motherboard regulates the functions of the rest of the hardware inside the computer. The motherboard has a battery on it that keeps the computer's clock running.

• WHAT IS A BIOS?

Do you remember what I said earlier about none of these questions disturbing your computer. This one only (in addition to the technical questions at the end of this book) will be an exception.

It is something you will probably never have to deal with unless you are installing something new in your computer – such as a new drive or more memory. You know that screen that comes up when you first turn on your computer? The one with all that information on it (such as how large your hard drive is, how many CD-ROM drives you have? That information is what your BIOS says your computer has.

Let's say you have installed more memory in your computer. When you turn on your computer, it's not going to continue as it normally does until you go into the BIOS (which is usually done by pressing a key like "F6" on your computer when it first starts up) and save it.

• WHAT IS A PROCESSOR?

The Processor – The brains of your computer. As far as hardware (hardware being the internal parts of the computer) this tells your computer almost everything that it needs to know to function properly.

• THERE ARE SO MANY NUMBERS. HOW DO I RECOGNIZE HOW FAST MY PROCESSOR IS?

First, note that this number (how fast your processor is) is always going to be in gigahertz if it's a new computer. When someone asks you the question, "How fast is my computer?" The answer will be something like 1.3 gigahertz or 2.0 gigahertz or 3.2 gigahertz, for example.

• WHAT DOES FRONT SIDE BUS MEAN?

There are many things on a motherboard that are all about the same thing – speed of processing information. The faster the front side bus, the faster information moves. For example, an 800 mhz front side bus is faster than a 400.

• WHAT IS CACHE?

It is pronounced "cash". Cache is a certain amount of extra memory. This is often called quick memory. The more cache on the motherboard, the quicker your computer operates.

• HOW MUCH IS A BYTE OF INFORMATION?

A byte is an amount of information. If you were to type the word "megabyte", you just typed a byte of information. A byte is a word or sequence of numbers that is about seven characters long.

• HOW MUCH IS A MEGABYTE OF INFORMATION?

A megabyte is a unit of measurement for computer memory or information storage. A megabyte is "one million bytes"[1]. If you type one word, you will have approximately the amount of info in a byte. OK, now times that byte by one million and you have a megabyte.

---

[1] The New Lexicon Webster's Dictionary of the English Language. Encyclopedic Edition, . Lexicon Publications, Inc. New York: 1989.

• HOW MUCH INFO IN A GIGABYTE?

A gigabyte is "one billion bytes"[2].

• I'D REALLY LIKE TO KNOW WHAT A DRIVE IS.

I know you're not going out to the country for a drive. You don't have to go anywhere, ladies and gentlemen. A drive is an area where you place, store, or manipulate information for immediate or later use.

• MY COMPUTER HAS A FAN IN IT?

Don't expect this fan to cool you down during the hotter months. This is that hum you hear when you push your computer's on button. This object is just what it says – a fan. This hooks on top of the processor and keeps the motherboard and the inside of the computer a certain cool temperature. If you didn't have this, you'd have a fire quickly.

• WHAT IS THIS ALPHABET ON MY COMPUTER?

Computers have different drives in them. Including:

Drive "A" – Your floppy disk drive.

Drive "B" – Most computers no longer have a drive "B".

Drive "C" – Usually your hard disk drive)

Drive "D" – Either this can be a hard drive, or it might be your CD-ROM drive.

Drive "E" – Either this can be another hard drive, or an additional CD-ROM drive.

Drive "M" – Some computers have this letter designated as a hard drive.

[2]. The New Lexicon Webster's Dictionary of the English Language. Encyclopedic Edition. Lexicon Publications, Inc. New York: 1989.

• WHAT'S THIS ITEM CALLED A FLOPPY DISK?

Floppy disk – A 3 ½" x 3 ½" object that stores 1.44 megabytes of information. Your floppy disk drive is usually denoted as *drive A*.

• I HAVE SEEN SOME DISKS, BUT THEY ARE NOT "FLOPPY" AT ALL.

The phrase floppy disk came about from the type of disk that came before the 3 ½" disk. This disk was 5 ¼", was thinner and more flexible. This disk only held 720K of information instead of the current 3 ½" disk that holds 1.44 megabytes of information.

• WHAT IS RAM?

Memory (or RAM) as it's commonly called. RAM stands for Random Access Memory. This is part of the computer's "memory", just as you and I have a memory. It is how much room it has to store or process knowledge, or more correctly, information. RAM typically comes in megabytes.

• I'M AT A COMPUTER SHOW AND I WANT TO BUY THE BEST / FASTEST MEMORY POSSIBLE? WHAT DO I BUY?

Make sure you ask the seller for "premium" memory. The greater the number of chips on the memory module, the faster it's going to go.

• WHAT IS A HARD DRIVE / HARD DISK / FIXED DISK DRIVE?

Hard Drive – Also called a fixed disk drive or a hard disk, a hard drive (sometimes abbreviated HD or FDD) is where all the information and computer programs [software] are stored on your computer. It is a rectangular device inside the computer with a lot of space to store information. Hard drives usually come in gigabytes (i.e. a 40 gigabyte hard drive). A hard drive (if you have only one – or the first one if you have more than one) is always denoted as "Drive C".

One more thing…not all hard drives have to be internal (meaning inside the computer). You can have an internal hard drive and an external one (meaning an enclosed hard drive connected to the computer from the outside. This is useful if you don't have any more room inside your computer for one.

• WHY DO THEY CALL IT A HARD DRIVE / HARD DISK?

It's called a hard drive because the round disk inside of the drive (that your information and computer programs are stored on) is hard, not floppy.

• THERE IS ANOTHER LITTLE GREEN/RED/YELLOW FLICKERING LIGHT ON MY COMPUTER. WHAT DOES IT MEAN?

When you are starting up a computer program or game and this light starts up, it means your hard drive is working. I don't mean that it's just operational; I mean it's processing data.

• WHAT'S THE DIFFERENCE BETWEEN THE HARD DRIVES

THAT SAY "5400" AND THOSE THAT SAY "7200"?

That number indicates how many times a minute the disk goes around. In other words, if you have a "7200 RPM" hard drive, it means the disk inside of your hard drive spins around 7200 times. Wow!

• IS HAVING YOUR HARD DRIVE "CRASH" A BAD THING?

Yes, ma'am or sir, it is. If you hard drive crashes – it means it is no longer useable.

• WHAT IS A MENU?

A menu is a list of options.

• WHAT'S THE DEFINITION OF A FILE?

Files are what floppy disks are made to store. A file is something you create in a computer program that contains information (text) or pictures that you want to store and retrieve later.

• I'M NOT GONNA GO OUT TO MY LOCAL OFFICE SUPPLY STORE FOR A FOLDER.

A folder is a place where many files are kept together.

• OKAY — NOW THAT I KNOW WHAT A FLOPPY DISK IS, HOW DO I FORMAT IT?

1. Click the "Start" button
2. Click "My Computer"

3. You should see a little disk "(Drive A)". Highlight this by left clicking on your mouse one time. Now, leave you mouse in the same place and right click. You should see a menu that gives you the option to format.

4. You should see "Start" and "Cancel" buttons. Click "Start" once.

5. When the little green bar is all the way over to the right. You're done.

6. Click "OK". Your floppy disk is now formatted.

• WHAT IS WORD PROCESSING / A WORD PROCESSING PROGRAM?

Word processing is typing and editing information. A word processing program is the typing program that you use to type your information.

• WHAT IS A CURSOR?

The Microsoft® Word® Windows®-based application is a place where you might see a cursor. You know that little blinking vertical line you see on your screen. That is a cursor. Wherever the cursor is when you start typing, that's where the letters will come up.

• WHAT IS A VIDEO CARD?

Video Card – This allows the computer to transmit visual images to your computer screen. Most video cards today are able to produce 3D characters and images on your computer. The video card plugs into a slot and comes out the back of the computer where it can be hooked via a cable to your computer monitor. Most video cards today have 128 megabytes of memory on them, so they can process information very quickly.

• TELL ME MORE ABOUT VIDEO CARDS.

Most video cards today are capable of producing realistic graphics in thousands of colors. Let me give you a little historical perspective – The early video cards only reproduced four colors. From four to thousands of colors…pretty nifty, huh?

• WHAT IS A VIDEO CAPTURE CARD?

It's a card that will let you capture video from your TV or movie camera and record it to a file on your computer.

• WHAT IS "TV OUT"?

"TV Out" is a jack on the back of some video cards, so you can see the output from your computer screen on your television.

• WHAT IS A PROMPT?

A prompt asks you for information by using words (i. e. "next" or "OK", etc), or a right arrow/blinking cursor like this (>, l, _), i.e. a MICROSOFT® MS-DOS® operating system command prompt. See the question, "WHAT IS A MICROSOFT® MS-DOS® OPERATING SYSTEM COMMAND PROMPT?" toward the end of this book.

• WHAT IS A MONITOR / FLAT SCREEN MONITOR?

Computer Monitor – The monitor is NOT the computer tower, it's the TV screen. It is where the information on the computer is presented to you. Until a few years ago, most computer screens were CRT's (Cathode Ray Tube). Most computer screens today are flat screen monitors, meaning they take up about a 5th of the space that computer monitors took up 20 years ago.

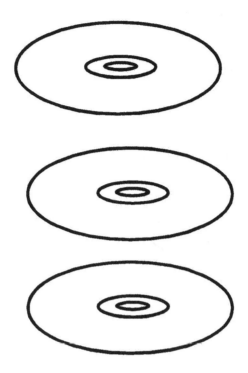

• AND NOW, GOOD PEOPLE, THE DEFINITION OF A CD-ROM...

A CD-ROM (Compact Disk Read Only Memory) – CD-ROM's are the media that most computer programs that you buy today come on. It holds many times the information that a floppy disk can hold. This kind of disk can only be read from, it can't have information saved on it once it has been finalized (see the chapter on burning CD's for more on this). Please note that a CD-ROM and a CD-ROM drive are two different things.

• I'M TIRED OF WRITING WITH MY PEN ON MY CD-ROM'S.
ISN'T THERE A WAY TO MAKE A TYPE-WRITTEN LABEL?

Yes. You can buy software that will help you make labels.

• WHAT IS A SPINDLE OF CD-ROMS?

It's just a container to keep them safe until they can be used.

• WHAT IS A CD-ROM DRIVE?

CD-ROM Drive – This drive is what holds those objects that we all hear so much about. It uses a laser to read the CD-ROM. You can, much like an external hard drive, buy an external CD-ROM drive.

• WHAT IS A MINI DISC?

Push the "open" button on your CD-ROM drive. Look closely at it. You see that little grove in the center. A mini disc goes there. It is read by your CD-ROM in the same way a standard size one is. It is just a smaller CD-ROM.

• WHAT IS A JEWEL CASE?

A jewel case is a hard plastic container that keeps your CD-ROM disk safe from dust and from being scratched.

• WILL MY COMPUTER RUN WITHOUT A HARD DRIVE?

Without a hard drive, memory, and a video card, it's almost impossible to get a computer to run properly.

• WHAT IS A KEYBOARD?

It's what you use to type information into your computer.

• WHAT ARE KEYSTROKES?

Each press of a key is called a keystroke.

• WHAT ARE ALL THESE EXTRA KEYS ON MY KEYBOARD?

On the top of some keyboards there are buttons (that are different from keyboard to keyboard that are internet buttons (see the chapter on the internet for more information. Not all keyboards have this).

Okay, starting from the top left, I'm going to describe the keyboard keys for you:

"Escape" Key (ESC): If you press this button, it will back you up or take you out of whatever program you are in. For example, if you press this during a game, it may bring up a menu or give you the option to stop the program right away.

"F1 – F12" are called the "function" keys. They are shortcuts to certain things, such as bringing up the option to save a document by only pressing a couple of keys.

Their functions depend what you are doing on the computer and what program you are in.

Keep going over to the right and you have three keys:

"Print Screen"

Prints or duplicates the information currently on the screen.

"Scroll Lock"

Makes it so the keys on the numeric keypad will not give you numbers when pushed, they will move you around your screen instead. Pushing this button again will turn it off.

"Break" key

It's usually used along with another key (such as "Control [CTRL])", to stop something that is currently happening from continuing. For example, if you were to format a disk under the command prompt area of the Microsoft® Windows® XP operating system and you pressed control and break at the same time, it would stop formatting the disk and bring you out to a command prompt. Now you can do something else. See the chapter on the Microsoft® MS-DOS® operating system later on in this book if you are confused.

Okay, now move over to the typewriter portion of the keyboard. Besides "A"-"Z", "1"-"0" and the standard punctuation marks, there are some other keys:

"Backspace" (or an arrow pointing to the left) lets you go back one space at a time.

• IS IT POSSIBLE TO USE THE "ENTER" OR "RETURN" KEY TO HIT THE "OK" BUTTON INSTEAD OF TIRING OUT MY HANDS?

Absolutely. If your hands are tired of moving that mouse, you can always tab to the "OK" button on your screen and press "Enter". This will have the same effect as using your mouse to click on the "OK" on the screen. Sometimes you won't even have to tab over. See the question, "WHAT IS A MOUSE?" for more help.

"Tab" moves your cursor (that blinking vertical line) about 5 spaces to the right every time you press it. How far it moves it depends on the settings. If you have it set to move five spaces, it moves five spaces. If you have it set to move seven spaces, it moves seven spaces.

"Enter" is like the "return" key on your typewriter. It moves you downward one or more lines at a time in a document.

"Caps Lock" makes it so you type nothing but capital letters or the punctuation marks above the numbers at the top of your keyboard.

"Shift" makes it so you type all capital letters, just the punctuation marks above the numbers at the top of your keyboard or can be used with other keys to do certain things (such as "Shift" + "Tab" moving you to the left vs. "Tab" itself moving you to the right, as talked about above.

MOVING DOWN TO THE BOTTOM OF THE KEYBOARD:

The *"Microsoft® Windows® logo®" button* is right next to the "Control" key "[CTRL]". This serves the same function as hitting the "Start" button with your mouse. NOTE: You can hit this button and use the arrow keys (at the right side of your keyboard) to move around without using your mouse.

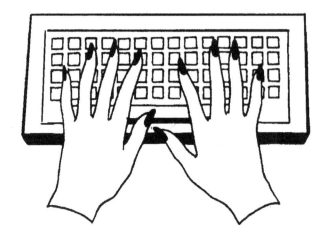

The "Alt" key: Usually pressed at the same time with other buttons to make the button(s) have an alternate function. Pressing "Alt" while in a Windows®-based business application will highlight the menus, so you don't have to use your mouse to access them, you can move around using the arrow keys.

The "Control" key "[CTRL]": Pretty much the same as the "Alt" key. This one is used in conjunction with other keys to do certain things (such as the "Control Alt Delete" key combination).

The "Menu" key: Your keyboard might have a key with a little menu and arrow on it. This is like pressing the right mouse button on your mouse. It brings up menus. The menus differ depending on what program you are in.

• WHAT DOES PRESSING "CONTROL ALT DELETE" DO?

A few things. Originally, in earlier versions of the Microsoft® Windows® operating system, whenever you pressed this combination once, it would restart your computer – no matter what you were doing.

Once you are in the Microsoft® Windows® XP operating system, if you press this combination once, you will get a screen with a list of Microsoft® Windows®-based applications, telling you whether they are running correctly or not. If a Microsoft®

Windows®-based application is frozen (not working), you can close it without restarting your computer.

If you press this key combination twice, the Microsoft® Windows® XP operating system restarts.

Moving over to the right of the computer keyboard, we have six keys. These keys are for moving around in certain Microsoft® Windows®-based applications (especially word processing):

The "Insert" key

Press this button and put your mouse cursor where you want to put in new words and, as you type the new words, the words next to them will be pushed aside. Press this button a second time to turn the function off.

The "Home" key

Moves you to the beginning of something (such as a page), or pressing "Control [CTRL]" and "Home" at the same time will move you to the beginning of the entire document.

The "End" key

Moves you to the end of something (such as a page), a sentence, or pressing "Control [CTRL]" and "End" at the same time will move you to the end of the entire document.

The "Page Up" key

Moves the page up, one at a time.

The "Page Down" key

Moves the page down one at a time.

The "Delete" key

Put your mouse cursor in front of a word you want to delete and press this button. It will delete the unwanted letters one by one.

The Arrow keys

Below the "End" button, we have four arrow keys. They move the cursor in the same direction as the arrow (up, down, left or right).

• THE "NUMERIC KEYPAD" AND THE "NUMBER LOCK"

On the far right side of your computer keyboard is the "numeric keypad". When you press the buttons on this keypad, you will type numbers, just like the other numbers on your keyboard. (NOTE: If you want to type numbers, the "Number Lock" must be on. This is the button that says "Num Lock". It's usually in the corner of the "numeric keypad". You can tell this button is in the "on" position when the light above it is on.

• DOES IT MATTER IF MY "NUMBER LOCK" IS ON OR NOT?

If it is, what will happen is whatever other function the keys serve (such as arrow keys, "Home" and "End" functions, "Page Up" and "Page Down", "Insert" or "Delete", or "Enter" functions) will be what happens when you press these buttons — instead of getting numbers.

• WHAT IS THE DIFFERENCE BETWEEN THE "NUMERIC KEYPAD" AND THE NUMBERS AT THE TOP OF THE

KEYBOARD?

Nothing. If you are using the calculator, for example, then you might find the "numeric keypad" easier to use.

• WHAT DOES THE "/ " KEY MEAN ON THE "NUMERIC KEYPAD"?

It is the key you use for division.

• WHAT DOES THE " * " MEAN ON THE "NUMERIC KEYPAD"?

It is the key you use for multiplication.

• WHAT DOES "+" MEAN ON THE "NUMERIC KEYPAD"?

It is the key you use for addition.

• WHAT DOES THE "-" MEAN ON THE "NUMERIC KEYPAD"?

It is the key you use for subtraction.

• MY KEYBOARD IS SITTING TOO FLAT. IS THERE ANY WAY TO PROP IT UP A LITTLE BIT?

Most people don't notice that there are two little stands on the bottom of your keyboard. Lift up your keyboard and you'll see what I mean.

• WHAT IS A MOUSE?

Remember to feed it cheese twice a day. Just kidding. A mouse is a hand control device that attaches to your computer. It is used to move that little arrow you see on your screen around, so you can get the computer to do things. Traditionally mice had to have a cord to work — these days you can find some cordless ones and, the latest evolution is the optical cordless mouse. Instead of using a ball to slide around your desk, it's battery powered and uses a beam of light to determine where the arrow on your screen goes. A mouse button can be clicked once or double clicked; depending on which function you want to perform.

• WHAT IS A MOUSE PAD?

It is a rectangular very thin item. It is used as a surface for sliding your mouse around so it moves freely.

• WHAT IS A SERIAL PORT?

It's that round plug you put your mouse into. It's just another way of saying "mouse port".

• WHAT DOES THE LEFT MOUSE BUTTON DO?

The left mouse button selects things. It selects things such as which Microsoft® Windows®-based application you want to use.

• WHAT DOES THE RIGHT MOUSE BUTTON DO?

The right mouse button presents various menu options, depending on the Microsoft® Windows®-based application you are in.

• WHAT IS "RIGHT CLICKING" ON SOMETHING?

"Right clicking" on something is holding your mouse arrow over an object and pressing your right mouse button.

• WHEN I PUSH THE RIGHT MOUSE BUTTON ON MY MOUSE, LITTLE MENUS COME UP. WHAT ARE THEY?

The right mouse button will bring up extra options for you to use with your Microsoft® Windows®-based applications. The specific options depend on what is on your screen. If, for example, you have the Microsoft® Windows® operating system desktop on your screen, you will get the options for arranging your icons automatically or creating a new folder.

• WHAT DOES THAT LITTLE WHEEL ON THE TOP OF MY MOUSE DO?

You can move it toward you to move down in your document. You can move it away from you to move up in your document.

• WHAT IS USB?

USB stands for Universal Serial Bus. USB devices (such as new mice and printers) deliver the information from the computer to the printer faster, and with more quality, than their older cousins, parallel ports (Those used about twenty-five pins and were a little over an inch across). Look on the back of your computer. Both of them are there.

• WHAT IS A USB PORT?

A USB port is where you plug in a USB device. This is usually a plug about a quarter of an inch tall located on the back of the computer.

• WHAT IS A USB KEY?

A USB key is a key you can stick into your USB port and save your files to it. This is useful for those who travel a lot or move documents between their computer at home and their computer at work. It is an update of the floppy disk.

• WHAT IS A USB HUB?

A USB hub is kind of like an extension cord. You plug the hub into a USB port on your computer. A hub has several connectors in case you have a computer that doesn't have enough free USB ports.

• WHAT IS A MODEM?

The most common type of modem in use today is the 56K modem (for a dial up connection). You connect your phone line to this and it connects you to the internet. You can also buy a cable modem, which connects you much faster.

• WHAT IS KPBS?

Kilobytes per second – It's how fast your dial up connection is on your modem.

• MY MODEM HAS TWO PHONE JACKS IN THE BACK OF IT. WHY?

This is useful if you do not want to keep plugging the phone cord into your computer's modem all the time. You take the cord coming from the phone jack on the wall and plug it into one of the jacks. Into the other jack, you plug one end into it and the other end into the phone.

You can now log onto the internet (see chapter on the internet later in this book), without having to plug or unplug any phone cords.

• IF IT'S A 56K MODEM, HOW COME IT NEVER CONNECTS AT 56K?

It depends on several factors – too many to list here.

• WHAT IS THE DIFFERENCE BETWEEN A REGULAR MODEM CONNECTION AND A CABLE MODEM CONNECTION?

Simply put, a 56K modem is the slowest connection and a cable modem connection is faster.

• WHAT IS A DVD?

It means Digital Video Disc (or Disk). It is usually used for movies. It is the successor to videotape. It also holds much more information than videotape.

• WHAT IS A DVD-ROM?

It's a lot like a CD-ROM. It can only have information put on it once. The biggest difference is that DVD-ROM's hold about 4.7 gigabytes of information. That's much more than a CD-ROM.

• IS A CD/CD-ROM/DVD/DVD-ROM BREAKABLE OR SCRATCHABLE?

Yes, if you bend it far enough. Under normal circumstances, it will not break, but it will scratch. The more scratch-free you can keep your disks, the better.

• TELL ME A LITTLE ABOUT DVD-ROM DRIVES.

These drives are a relatively new invention for computers, just like they are for your television. It's a CD-ROM drive, but has the capability to play your DVD movies also.

• CAN I PLAY DVD MOVIES ON MY COMPUTER?

Yes! If you have a DVD-ROM drive, of course you can, just remember to turn on your speakers so you can hear it.

• WHAT IS A SCANNER?

Scanners are devices to get graphics (pictures) into your computer. Every device has its "measuring stick". For scanners, it's something called DPI (Dots Per Inch). The more dpi you scan a picture at, the better it scans into the computer.
These days, scanners can be hooked up to your computer through your printer to function as a copier.

• WHAT IS A PRINTER?

A printer produces on paper what you have created on the computer. There are two types of printers that are widely being used today: inkjet printers and laser printers.

Inkjet printers are the more economical type of printer. For most people's needs, these produce adequate results.

The other type of printer is a laser printer. It uses a laser (hence the name) to print on the paper. It is about 10 times faster, 5 times quieter, and some claim they produce much sharper results. This printer is more expensive than an inkjet, but would be worth it if you had a business, for example.

• WHY DO I NEED TWO CARTRIDGES FOR MY PRINTER (DUAL CARTRIDGE PRINTERS)?

Most new printers use two cartridges. One has nothing but black ink in it. The other has three different colors of ink in it to make up all the other colors of the rainbow.

• WHAT IS A RADIO BUTTON?

A radio button allows you to choose a particular option or choice on an electronic form. It's a little round area that you click on with your mouse. See the next question on how to conserve printer ink for an example of a radio button.

• MY INK CARTRIDGE IS LOW ON INK. IS THERE A WAY TO CONSERVE INK?

Yes, absolutely.

1. Go to "File" and click "Print".

2. Click "Properties"

3. On the left side, you will see three radio buttons. One says "Draft", one says "Normal", and the other says "Best".

4. With your mouse, click "Draft".

5. Click "OK".

6. This will take you back to where you were before. If you still want to print your document, click "Print".

• I PRINTED SOMETHING THAT WAS CUT OFF AT THE RIGHT SIDE OF THE PAGE.

1. Go to "File" and click "Print".

2. Click the "Features" tab

3. On the left side, you will see two radio buttons. One says "Portrait" (which is where it was set when you started), and one says "Landscape".

4. With your mouse, click "Landscape".

5. Click "OK".

6. This will take you back to where you were before. If you still want to print your document, click "Print".

• IN A PREVIOUS QUESTION, YOU MENTIONED SOMETHING ABOUT A DIFFERENT KIND OF "TAB" THAN THE ONE ON THE KEYBOARD. WHAT IS THIS KIND OF "TAB"?

You're right. In the question about "Landscape" printing, I mentioned a different kind of tab. This kind of tab refers to the little half-circle items that you will often see in the Microsoft® Windows® operating system. Let me give you an example:

1. Go to the "My Computer" icon.
2. Right click on it with your mouse.
3. Select "Properties"
4. See where it says "General" and "Shortcut" at the top? Those are called "tabs".

• I DECIDED THAT I DIDN'T WANT TO PRINT SOMETHING AND IT'S ABOUT TO PRINT.

In this particular situation, the first thing to do is turn off your printer, then you can cancel what's printing by:

1. Double clicking on the little printer icon that you will see right after you tell the computer to start printing something.
2. You will see a "box" with some menus.
3. Select the "Printer" menu.
4. Select "Cancel All Documents".

• WHAT IS A PAPER JAM AND WHY DOES IT ALWAYS SEEM TO HAPPEN AT TO ME AT THE WRONG TIME?

A paper jam is when too much paper gets into the workings of your printer at once, and it clogs up your printer. Make sure your paper doesn't stick together. If you just put it into your printer, it may be sticking together and that is what is causing your paper jam. As to why it always seems to happen at the wrong time, my advice is: don't print something in a hurry.

CHAPTER 3 – NAVIGATING THE MICROSOFT® WINDOWS® XP OPERATING SYSTEM

Every computer needs an operating system to run on. As of this writing, the Microsoft® Windows® XP operating system is the current version.

• WHAT DOES INSTALL MEAN?

To install a Microsoft® Windows®-based application is to put in onto your computer so you can use it (see "WHAT IS THIS ALPHABET ON MY COMPUTER?" before you do this).

• WHAT IS THE MICROSOFT® WINDOWS® OPERATING SYSTEM DESKTOP?

It is the main area that holds your screen icons.

• WHAT IS AN ICON?

An icon is a little picture on your screen that is a shortcut to a Microsoft® Windows®-based application that's on your computer.

• HOW DO I BEGIN USING THE MICROSOFT® WINDOWS® XP OPERATING SYSTEM?

Need help figuring out the Microsoft® Windows® XP operating system? No problem! It all begins with the "Start" button. The "Start" button is located in the lower left hand corner of the screen. From here, you can access almost all of your computer functions.

• WHAT IS THE MICROSOFT® WINDOWS® XP OPERATING SYSTEM TASKBAR?

This is the blue bar at the BOTTOM of your screen. It includes the "Start" button. It also shows the Microsoft® Windows®-based applications and/or files that are currently in use, and some icons for quick access to what's on your computer. To open one of the items listed, click with your left mouse button on the Microsoft® Windows® operating system taskbar and then click on the item you want to open. The clock is in the lower right hand corner.

• I DON'T WANT THE MICROSOFT® WINDOWS® OPERATING SYSTEM TASKBAR AT THE BOTTOM OF MY SCREEN. WHAT DO I DO TO TEMPORARILY GET RID OF IT?

1. Click on the "Start" button.
2. Click "Control Panel"
3. Double click "Taskbar and Start Menu".
4. With your mouse, check the box that says, "Auto-hide the taskbar".
5. The only time that the Microsoft® Windows® operating system taskbar will appear now is when you move your mouse arrow to the bottom of the screen.

• WHAT IS THAT HOURGLASS FOR?

When you see that little hourglass, it means the computer wants you to wait for whatever information that is being processed to finish — before you do something else.

• WHAT IS THAT LITTLE ARROW IN THE LOWER RIGHT HAND CORNER?

If you click on it with your mouse, it will show you the rest of the icons — because the icons have been hidden so you have room for things.

• WHY DO THEY CALL IT THE MICROSOFT® WINDOWS® OPERATING SYSTEM AND DO I HAVE TO CLOSE ONE MICROSOFT® WINDOWS®-BASED APPLICATION BEFORE I CAN OPEN ANOTHER?

Each Microsoft® Windows®-based application that you run opens into a different screen. If you have several things running at once, you don't have to close one to get to the other. Check the Microsoft® Windows® operating system taskbar (at the bottom of the screen) to see what's open and you can switch between Microsoft® Windows®-based applications very easily.

• WHAT IS A DROPDOWN MENU?

There are two kinds of dropdown menus:

The first kind provides choices to select a different folder or drive to view, for example. It usually has a little arrow (pointing down) to one side of it. To see an example of this, do the following:

1. Press the "Start" button
2. Select "Run"
3. Select "Browse"
4. Click the down arrow next to where it says, "Look in". What you see next is a dropdown menu.

The second kind of dropdown menu can be found in any Microsoft® Windows-based business application. They are also called dropdown menus because they drop down when you click with your mouse on such words as "File", "Edit", "View", etc., which are located at the top of the screen. See the questions, "I WANT TO SAVE OR VIEW A DOCUMENT ON ANOTHER DRIVE. HOW DO I CHANGE THE DRIVE OR FOLDER I AM LOOKING AT?" AND "I NEED TO INSTALL A NEW COMPUTER PROGRAM ONTO MY COMPUTER. WHAT DO I DO?" for more info.

• I WANT TO SAVE OR VIEW A DOCUMENT ON ANOTHER DRIVE. HOW DO I CHANGE THE DRIVE OR FOLDER I AM LOOKING AT?

You will probably see something that says "Local Disk (C:)". To change the drive you are looking at:

1. Click the down arrow with your mouse and you will see a list of other drives (a dropdown menu).
2. Click on one of those other drives and click "OK" (or just double click on the other drive) and you will see its contents.
3. If you need to see inside a folder on that particular drive, double click on it.

• I NEED TO INSTALL A NEW COMPUTER PROGRAM ONTO MY COMPUTER. WHAT DO I DO?

Put your CD-ROM into the drive and wait for the light to go off. Now, do the following:

1. Click the "Start" button
2. Click "Run"

3. Click "Browse"

4. There is a dropdown menu next to where it says, "Look in".

5. With your mouse arrow, click here and select "Drive D", "E", or even sometimes "M", for the letter of your CD-ROM drive.

6. Browse until you see "setup" or "setup.exe".

7. Double click with your mouse or click on it once and then click "OK". You can then follow the prompts on your screen to install it on your computer.

• THE MICROSOFT® WINDOWS®-BASED APPLICATIONS LISTED ON MY COMPUTER ARE OUT OF ORDER. IS THERE ANY WAY TO ALPHABETIZE THEM?

Yes, there is.

1. Click the "Start" button.

2. "Programs"

3. Hold your mouse arrow over one of the words on the list and right click with your mouse. You will see the word "sort" on the menu.

4. Select "sort" and the items on your list will be alphabetically sorted.

• WHAT IS THE *RECYCLE BIN*?

Look in the lower right hand corner of your screen. You should see something called the *Recycle Bin*. This is how you dispose of files that you no longer want on your hard drive.

• HOW DO I EMPTY THE *RECYCLE BIN*?

You can do this in one of two ways:

1. Put the mouse arrow over the *Recycle Bin*.

2. Press the right most button on your mouse.

3. Click (with your left mouse button) on the option that says "Empty *Recycle Bin*".

4. When it asks you if you are sure, select the "Yes" button.

OR

1. Double click (using your mouse's left button), on the *Recycle Bin*.

2. When the screen pops up, click on "File".

3. Click on "Empty *Recycle Bin*".

4. When it asks you if you are sure, select the "Yes" button.

• I DELETED SOME FILES FROM MY COMPUTER, BUT THE SPACE ON MY HARD DRIVE IS STILL TAKEN UP? WHAT'S WRONG?

Take another look and make sure you emptied the *Recycle Bin*.

• WHAT DOES "DEFAULT DRIVE, FOLDER, OR BUTTON" MEAN?

To default, in computer language, is not a bad thing. A default drive, folder or button is the one that the Microsoft® Windows® operating system will "come up to" first. An example of a default drive would be drive "C": (your first hard drive) because the Microsoft® Windows® operating system boots up to it first.

An example of a default folder is this: if you went to save a document, the default folder would be *"My Documents"*.

An example of a default button might be "Save" or "OK".

• WHAT ARE THOSE THINGS IN THE BLUE BAR IN THE TOP RIGHT HAND CORNER OF MY SCREEN?

There are three:

The one that looks like a little line will take the screen you opened up most recently and "minimize" it (shrink it down). You can "maximize" (or make it big again) by clicking on the "title" that will now be in the Microsoft® Windows® operating system taskbar. If you have several (meaning more than three or four) items open at once, they will consolidate into one button. All you have to do in this case is click on the button and it will present you with a list with all your open Microsoft® Windows®-based applications. Just continue to hold down your left mouse button and drag the cursor (arrow) to the program (or file) that you wish to view.

The one in the center has two boxes in it. If you have a full screen and you click this button, the full screen will become a smaller screen, and if you click it again, it will return to a full screen.

The third one (the one in the far right top corner with a red "X", will close the current open screen.

• WHEN I STOP MY MOUSE ARROW OVER SOMETHING, IT SHOWS LITTLE BOXES WITH WORDS IN THEM. WHAT DOES THIS MEAN?

These are different from the options discussed in the question about right clicking. You don't have to right click to do this. This is something nifty in the Microsoft® Windows® XP operating system. If you hold your mouse arrow over a certain button, icon, etc. in the Microsoft® Windows® XP operating system, you will get more information on about that item. For example, if you hold your mouse arrow over the clock (time) in the bottom right corner of the screen, you get a little box that tells you the day and date as well. This also includes the buttons at the top of the screen in many Microsoft® Windows®-based applications.

• I'M IN THE MICROSOFT® WINDOWS® XP OPERATING SYSTEM AND I'VE INSTALLED A MICROSOFT® WINDOWS®-

BASED APPLICATION ON MY COMPUTER. NOW HOW DO I USE IT?

One of two ways. If you have an icon (little picture) for the program on your computer, you can double click on it (click twice quickly) and it will start your Microsoft® Windows®-based application. Or to do it the hard way:

1. Click "Start"
2. Click "Programs"
3. Use the mouse to find the name of the program and a side menu will come out.
4. Click the name of the program from the side menu and your program will start up.

Depending on the program, you must have the CD in the CD-ROM drive for the program to run, because games and similar applications usually get the audio from the CD-ROM itself.

• I WANT TO FIND A FILE ON THE COMPUTER. HOW DO I DO THAT?

There is more than one way to do this. You can use "Search":

1. Click the "Start" button.
2. Click "Search"
3. With your mouse, left click on "All Files and Folders".
4. In the top of the two boxes, type the name of the file and click "Search".
5. You will see a list of files. Double click on the one you want and it will open.

Another way to do this is to use "My Computer":

• WHAT IS "MY COMPUTER"?

"My Computer" shows all the drives on your computer. You can proceed to "My Computer" by:

1. Clicking "Start"
2. Click "My Computer"
3. It will show you all the drives on your computer. If you just saved a file to your floppy disk, you would find it on drive "A", for example. Therefore, you would click on drive "A" and your computer would show you what is on that disk.

• WHAT IS "MY DOCUMENTS"?

Now a word about the *"My Documents"* folder. This is a folder where (by default) most of your new files that you make are saved. You get to *"My Documents"* by clicking "Start", and *"My Documents"*. Then something will come up that says, "Files stored on this computer". Below it will *say "My Documents"* or something similar; this is determined when you set up the Microsoft® Windows® XP operating system).

Or…you can use the *"Microsoft® Windows® Explorer®"* Windows®-based application to find files on your computer.

• WHAT IS THE *"MICROSOFT® WINDOWS® EXPLORER®"* WINDOWS®-BASED APPLICATION?

In the Microsoft® Windows® XP operating system, you get to this by:

1. Clicking "Start"
2. Clicking "Programs"
3. Go to "Accessories"
4. Click on the *"Microsoft® Windows® Explorer®"* Windows®-based application.

This lists options for the "My Computer" folder, the *"My Documents"* folder, and the files in the particular folder you are viewing, and all the drives on your computer all in one comprehensive screen.

• I DOUBLE CLICKED ON A FOLDER TO SEE WHAT WAS INSIDE IT. I WENT INTO THE WRONG FOLDER. WHAT CAN I DO?

In the left top corner, there is a button that says back. Each time you click on this, it will take you back one screen. Click on it until you are where you want to be.

• HOW CAN I CHECK TO SEE HOW MUCH SPACE IS TAKEN UP ON MY HARD DRIVE?

1. Go to the *"Microsoft® Windows® Explorer®"* Windows®-based application (Click the "Start button", "Programs", "Accessories", and then, click on the *"Microsoft® Windows® Explorer®"* Windows®-based application).
2. Click once on "My Computer" on the left side of the screen.
3. Right click on drive "C", "D" or whatever hard drive you want to check, and select "Properties" from the menu that comes up and you will be shown how much space is taken up, and how much is free.

• I'M IN *THE "MICROSOFT® WINDOWS® EXPLORER®"* WINDOWS®-BASED APPLICATION OR *"MY DOCUMENTS"*. I

RIGHT CLICK ON A FILE AND ONE OF THE MENU OPTIONS SAYS, "OPEN WITH". WHAT DOES "OPEN WITH"… DO?

The Microsoft® Windows® XP operating system generally assigns a particular program to open a particular type of file. For example, if you had a picture file, it opens with the same program every time. "Open with…" is there in case you want to open a file in a different program. To open a file in a different program:

1. Right click on the picture or file that you want to open in a different program. One of the options will say, "Open with"
2. With your left mouse button, click on "Open with"
3. You will be given a list of Microsoft® Windows®-based applications, which are installed on your computer, to open it with.
4. Choose the one you want and click once with your left mouse button.
5. If the Microsoft® Windows®-based application you want is not listed, click "choose program" and select one of the others listed.

• WHAT DOES DRAGGING AND DROPPING A FILE MEAN?

It's basically a quick way of copying a file.

• HOW DO I DRAG AND DROP A FILE?

1. Go into the *"Microsoft® Windows® Explorer®"* Windows®-based application.
2. Click on "My Computer" on the left side of the screen so you can see the drives on your computer.
3. Click and hold down the left mouse button on the file you want to copy.
4. Move your mouse arrow across the screen and over to the drive or folder where you want to put it and let go of the mouse button.
5. The file will now be copied to the new folder.

• I'M TRYING TO COPY FROM ONE FOLDER TO ANOTHER IN THE *"MICROSOFT® WINDOWS® EXPLORER®"* WINDOWS®-BASED APPLICATION. IS THERE A WAY I CAN SEE BOTH FOLDERS AT THE SAME TIME?

To more easily drag and drop a file from one folder to another you need to be able to see both folders. In the *"Microsoft® Windows® Explorer®"*

Windows®-based application, on the left side, is a list of the drives, folders, and files on your computer.

1. Go to the *"Microsoft® Windows® Explorer®"* Windows®-based application and click on "My Computer".
2. Click on the drive that contains the folders you want to see (drive "C", for example).
3. Click on the folder that contains the file or files you want to drag and drop.
4. If the other folder is not visible, repeat steps 1 and 2 so you can see it.
5. Drag and drop files as necessary. You will probably only be able to see inside one folder at a time (meaning it shows on the right hand part of your screen), so make sure it's the one you are coping files from.

If you can't need to see both at the same time and can't you can always copy and paste the file.

• HOW DO I COPY AND PASTE OR CUT AND PASTE A FILE?

Copying and pasting a file is to make a duplicate of the file and move it to a new location. Cutting and pasting is to completely move the file from one location to another.

1. Click once on the file you want to copy. This will highlight it.
2. Right click on it and select "Copy". You can also find this (and the "cut" option under the "Edit" menu.
3. Go to the drive or folder that you want to copy the file to.
4. On the right side of your screen, on the white part, right click (or if you want to put it in a certain folder, right click on top of the folder).
5. Select "Paste".
6. Your file has now been copied to the new location.

• WHAT IS SCREEN RESOLUTION?

Do you remember when you would just bang the daylights out of one of the television sets to get a better picture? Well, it's kind of like that – without the banging. Everything you see on your screen is made up of pixels (little dots), much like on your TV screen. Screen resolution is how many pixels you see on your computer screen. Common resolutions are 640 x 480, 800 x 600, 1024 x 768, and 1280 x 1024.

• I NEED TO CHANGE MY SCREEN RESOLUTION.

1. Click the "Start" button
2. Click on "Control Panel"
3. Double click on "Display"
4. Click on the "Settings" tab
5. There is a slider. The default resolution is 800 x 600.

6. To increase or decrease it, move the slider with your mouse left or right.

7. When it's where you want it, click on "Apply".

8. If it asks you whether you want to keep this new resolution, click "Yes" if you want to keep it (and "No" if you don't).

• WHAT DOES REFRESH RATE MEAN AND HOW DO I CHANGE IT?

It's how fast your computer picture re-produces itself. To change it:

1. Go to the "Control Panel" (Click "Start", then click "Control Panel").

2. Double click on "Display"

3. Go to "Settings"

4. Go to "Advanced"

5. Go to "Monitor" and you will see your choices.

6. Click on the little down arrow and it will show you a menu of your choices.

7. Click on the one you want and press "Apply".

8. Then, click "Yes" or "No".

• I WANT TO RENAME A FILE I HAVE ALREADY SAVED.

The easiest way to do this is to go into the program in which the file was created and do it that way, but there is a way to do this through the *"Microsoft® Windows® Explorer®"* Windows®-based application:

1. Click on the file with your left mouse button once to highlight it

2. Click on the center of the file name one more time with your left mouse button. You should now see a little box around the file name. You now have the option to rename the file.

3. Type the new file name and press "Enter". Voila, your file is renamed.

• WHAT IS THE "CONTROL PANEL"?

In the Microsoft® Windows® XP operating system, the "Control Panel" can be reached by:

1. Clicking "Start"
2. Click "Control Panel"

The "Control Panel" is like the center of your computer. The "Control Panel" includes, but is not limited to, the following:

"Accessibility Options"

This includes options for those who are visually challenged or hearing impaired. This includes shortcut keys (like setting it up so you can do a function that requires pressing multiple keys at the same time with one keystroke, or letting the Microsoft® Windows® XP operating system show you on the screen when it makes a warning sound.

WHAT IS A DRIVER?

A driver is a file that you install (usually when you install a new piece of hardware into your computer) to make a hardware device work.

"Add Hardware"

This helps you install new hardware and the drivers (files that make it work) to go with it. It searches your computer for new hardware and gives you the option to install it.

"Add or Remove Programs"

It presents you with a list of Microsoft® Windows®-based applications that are on your computer (along with how large they are and how often they are used), and gives you the option to uninstall them. It will tell you whether or not it removed all the computer files related to the program and whether you need to remove any of them yourself.

"Administrative Tools"

The "Administrative Tools" area references some of the things that are on the computer and what they do.

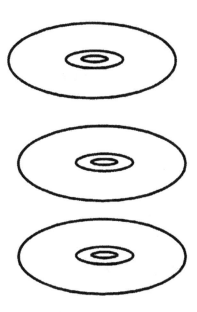

"Date and Time"

There are tabs here for "date and time" and "time zone", and a tab to synchronize your clock with an internet site.

"Display" tab

This controls several of your screen options. This includes tabs for the following:

"Themes" tab

Here you can choose (from a drop down menu), which set of colors/types/sounds you want for your computer – and such things as what sounds you hear when certain things happen.

"Desktop" tab

Here you can change the color of your Microsoft® Windows® operating system desktop, the wallpaper (pictures) on it and whether you want the picture centered, tiled or stretched.

"Screen Saver" tab

Here you can change your screen saver…

• WHAT IS A SCREEN SAVER?

A screen saver is something that comes up on your computer screen to prevent phosphorus burn. Phosphorous burn will burn an image on your computer screen if the image is left in the same place too long. The screen saver keeps moving to prevent this.

• HOW DO I CHANGE MY SCREEN SAVER?

1. Go to the "Control Panel".
2. Double click on "Display"
3. Click the "Screen Saver" tab
4. Click on the arrow that's pointing down to see the screen savers available.
5. Click on the one you want.
6. The screen saver appears if you leave your mouse and/or keyboard untouched for a certain number of minutes. If you want to change how soon it appears, click on the up or down arrows shown here to change that number.
7. When you are done, click "Apply". Your screen saver is now changed.

"Appearance" tab

Here you can change your color scheme.

"Settings" tab

Here you can change your screen resolution and the quality of your colors.

Now back to the other "Control Panel" options...

"Folder Options"

This asks you how you want your folders displayed. It asks if you want them displayed in the Microsoft® Windows® XP operating system format (which gives you little captions to tell you what everything does), or you can display them in the format that was commonly used in a previous versions of the Microsoft® Windows® operating system.

• WHAT IS A FONT?

A font is a style of type. You may also install new fonts on your computer (from a USB key, floppy disk, or CD-ROM).

"Game Controllers"

Lets you install and/or set up any game controllers that might be installed on your system.

"Internet Options"

OK, this area brings up your browser (which for the Windows® XP operating system is the *"Microsoft® Internet Explorer®"* browser). You can change many options to your own preferences. We're going to go through the tabs one at a time.

"General" tab

Includes the options to set your home page, deleting "temporary internet files" (see the chapter on the internet), and clearing your history (the list of places where you have visited on the internet).

"Security" tab

You can set your computer up to your own preferences here.

"Privacy" tab

Determine how much privacy you want. You can set how many types of cookies you will allow to be downloaded to your computer. Cookies are files from the websites you visit that help them work properly (see the question, "WHAT ARE COOKIES?").

"Content" tab

This tab has more options for your privacy.

"Connections" tab

Allows you to make new internet connections for your computer (meaning what Microsoft® Windows®-based applications, etc. your computer uses to connect to the internet), and to configure existing ones.

"Programs" tab

You may select items from a dropdown list that the Microsoft® Windows® XP operating system will use automatically for internet, E-mail, etc.

"Advanced" tab

Here you can select or de-select check boxes that control such things as whether your computer underlines web-links or whether your printer prints background colors or not).

Okay, that's it for the "Internet Options" area. On to the next "Control Panel" area...

"Keyboard"

Options to fine-tune your keyboard, such as the character repeat rate (how fast the characters come up on your screen when you hold a key down) and the cursor blink rate.

"Mouse"

Options to customize your mouse…

"Buttons" tab

Select whether you want to change the switch the functions of your left and right mouse buttons, how fast or slow the double click speed is, and an option that lets you hold down the mouse button without actually holding it down.

"Pointers" tab

Specify what your mouse pointers (that little arrow) looks like.

"Pointer Options" tab

More mouse pointer specifications here. There is a very useful option here that I suggest using. Check the box here that allows the mouse pointer to go automatically to the nearest default button, such as a button that says "OK".

"Wheel" tab

For computers that have a wheel mouse attached to them. It controls how many lines your mouse will scroll each time you turn it.

Next, we have...

"Network Connections"

Tells what network connections (i.e. internet, etc.) you have and whether they are currently in use or not.

"Phone and Modem Options"

Options here include being able to change the area code that you are dialing from.

"Power Options"

Options here include choices such as setting how long before your monitor turns off if you leave the computer alone.

"Printers and Faxes"

Add or delete printers and fax machines from your computer.

"Regional and Language Options"

Here you can set the language you want the Microsoft® Windows® XP operating system to appear in.

"Scanners and Cameras"

You can add scanners or digital cameras, or both, with this option.
Tip: When you turn on your digital camera it will show up in this area. If you double click on the camera, it will give you the option to transfer the pictures easily from your digital camera.

"Scheduled Tasks"

You can set up tasks here for the Microsoft® Windows® XP operating system to perform on a regular basis, such as running a certain Microsoft® Windows®-based application on the same day every week.

"Sounds and Audio Devices"

Options for your sound card, such as "volume control", "mute", and, under the "Sounds" tab, you can control the sounds you want to use for certain circumstances (i.e. what sound you hear when the Microsoft® Windows® XP operating system starts up).

"System"

A very useful area. There are several tabs included within this area, but we're only going to concentrate on two or three of them.

"General" tab

This tab tells you what version of the Microsoft® Windows® operating system you are using, who the computer software is registered to, the speed of your computer and how much memory you have installed in your computer.

• WHAT IS THE "DEVICE MANAGER" AND HOW DO I GET TO IT?

Under the "System" area of the "Control Panel", click the "Hardware tab" and then "Device Manager". It will show a list of all the different devices on your computer. If you double click on one of those device types, the computer will tell you how many and what kind of devices you have. After that, if you highlight a single device and right click on it, there is a menu. It will let you do such things as updating the device driver, to disable the item, uninstall the item, or you can scan the computer for hardware changes, meaning if you just installed some new hardware in your computer, the Microsoft® Windows® XP operating system will recognize the hardware and give you the option to install a driver for it.

Remember, to get here:

1. "Start" button
2. "Control Panel"
3. Double click on "System"
4. Click on the "Hardware" tab.
5. Click on "Device Manager".

"Taskbar and "Start" Menu"

Options to adjust the Microsoft® Windows® operating system taskbar or "Start" menu to your liking, such as the ability to lock it in place if you want to and switching between the newest "Start" menu and the "Start" menu from earlier versions of the Microsoft® Windows® operating system.

User Accounts

Here is where you can set up an account for each person using the computer (so each person can set up a different screen background or other personal preferences.

• HOW DO YOU CLOSE THE MICROSOFT® WINDOWS® XP OPERATING SYSTEM?

1. Click the "Start" button
2. Click "Turn off Computer"
3. Click the button that says, "Turn Off".

This will shut down your computer properly.

• WHAT HAPPENS IF I DON'T SHUT DOWN MY COMPUTER PROPERLY?

If you don't shut down your computer properly, you will probably damage some of the files on your computer (not necessarily the files you have written yourself, but the files responsible for letting your software run properly. What this boils down to is that you might have some problems with your software and you might have to re-install them, so keep the installation CD's handy.

• OKAY, I ACCIDENTALLY SHUT DOWN MY COMPUTER WRONG.

What your going to see is a diagnostic screen that is checking your computer files to see that they're okay. This is normal when you shut your computer down wrong. It will fix what it can and get your computer back to normal. The screen you're used to seeing when you turn your computer on will come up when this process has finished.

• WHAT DOES THE "LOG OFF" BUTTON DO?

The Microsoft® Windows® XP operating system is set up like many new cars. You know how they are set up so the seat is adjusted just right. The Microsoft® Windows® XP operating system is much the same way. The Log Off button does just what it says…it logs you (meaning, you personally) off the Microsoft® Windows® XP operating system. That way, someone else can log on with their own personal settings that are the same (or different) from yours, without rebooting. These can include different preferences for screen resolution, the picture on your screen, etc.

• WHAT DOES THE "STAND BY" BUTTON DO?

This button puts the computer in what is called "sleep mode". The point of this button is so you can "half way" shut your computer down. This serves two purposes – to save energy and to turn your computer on quickly if you leave and come back to it often.

• HOW DO I BRING MY COMPUTER OUT OF "SLEEP MODE"?

Move you mouse and/or press a button on your keyboard. When the screen comes up telling you how many Microsoft® Windows®-based applications you have running, click with your mouse arrow right there and you will see what was running when you put your computer in "sleep mode".

• WHAT DOES THE "RESTART" BUTTON DO?

Unfortunately, I cannot say this button is a lot like your car, because I think we'll have to wait a couple of hundred years before pressing a button can restart your car. Anyway, the "Restart" button shuts down the Microsoft® Windows®-based applications you have open, shuts down the computer, and restarts it.

• DO I NEED TO HAVE A SEPARATE ORIGINAL COPY OF THE MICROSOFT® WINDOWS® XP OPERATING SYSTEM FOR EACH COMPUTER I OWN?

Yes. You must buy a copy for each computer you have.

CHAPTER 4 – THE ENTERTAINMENT

If you are a music and movies person, read this chapter.

• WHAT DO I DO IF I DON'T KNOW WHAT MICROSOFT® WINDOWS®-BASED APPLICATION TO OPEN A PARTICULAR FILE WITH?

It's hard for me to advise on this one because no one person has the same programs on his or her computer – but I'll give it a shot.

If you want to open something you know is a picture, you need a program that you know opens pictures, such as the *"Microsoft® Internet Explorer®"* browser.

If you want a program to play a sound, song, or DVD, you might want to use something capable of playing music such as the Microsoft® Windows® Media® Player.

• WHAT IS THE MICROSOFT® MICROSOFT® WINDOWS® XP OPERATING SYSTEM MEDIA® PLAYER?

It is intended to play all the different types of files on your computer. Usually the most common file types it plays are the music and sound files and the DVD movies. To get to the Microsoft® Windows® XP operating system – do this:

1. Click the "Start" button.
2. Select "Programs"
3. Select "Accessories"
4. Select "Entertainment"
5. Select the Microsoft® Windows® XP operating system "Media® Player".

• I WANT TO OPEN A PARTICULAR FILE IN THE MICROSOFT®

WINDOWS® XP OPERATING SYSTEM MEDIA® PLAYER, AND I NEED TO FIND THE FILE MENU. WHERE IS IT?

As you have probably found out by now, the menus on most all Microsoft® Windows®-based applications are the same format, for the most part. From left to right, it's usually "File", "Edit", etc. Anyway, a "File" menu is included in here too.

1. There is a button on the top left that looks like an up arrow and a down arrow on top of each other. Press this and this will expand your view. The "File" menu will be located on the top left.

2. If you want to open something that is already listed for you, just double click on it with your mouse or press the "play" button.

3. If you want to open something that is not listed, click "File" and "Open", then select the file you want to open with your mouse.

CHAPTER 5 — BUYING A NEW COMPUTER

• UPGRADABILITY? TELL ME MORE.

When you buy a new computer, you want to make sure that you have room to upgrade it (add new stuff to it), without having to buy a new one. For example, when you decide on how much memory you want in it, be sure to ask whether there is room to upgrade the memory (add even more) later. You may also want to make sure that it has more drive bays than you need, just in case you want to add some new technology to it.

You want to buy a new computer. The question probably running through your mind right now is:

• HOW DO I KNOW WHAT TO GET ON MY NEW COMPUTER?

Well, computers are highly personal things to their owners. The options you have on your computer are no exception. Here are some recommendations:

Processor Speed (How fast)

Games are what usually push computers to go faster and faster, even more so, depending on the sophistication of the game. Many people like to use it for going on the internet, but if you're going to be playing a lot of games, you might want to get a computer that is slightly faster. The slowest most new computers run these days are 1.8 gigahertz (which is pretty quick!). The fastest ones are over 3 gigahertz (which is even quicker).

Monitor

You can buy a CRT (Cathode Ray Tube) monitor or a flat panel monitor. I recommend the latter.

Front Side Bus

800 mhz is what's on many computers these days

RAM (memory)

I recommend getting a computer that has at least 256 megabytes of RAM, if not 512. Check out the possibilities of being able to add more later, whether you need to add more or not.

Floppy Drive or No Floppy Drive

I do recommend the inclusion of a floppy "(A:)" drive. It lets you use older programs (ones that only came on a floppy disk).

Hard Drive Size

It depends on how many and how large the programs are that you are going to put on it. I would recommend that you get a hard drive that is 60 gigahertz or better and runs at 7200 RPM (as opposed to 5400).

Video Card

A video card capable of reproducing 3D graphics is necessary. In addition, it should have at least 128 MB of RAM on it.

Desktop or Tower Case?

Tower case.

DVD-ROM Drive or CD-ROM?

If you want to be able to play DVD's on your computer, you must go with a DVD-ROM drive. If you just want to run programs, a CD-ROM drive is all you need.

CD Burner?

Yes, if you want to be able to save files to a CD.

DVD Burner?

Yes, if you want to be able to save files to a DVD (remember, a DVD holds more information than a CD).

Modem

A 56K modem works just fine for most people's needs.

Cache

You want at least 256KB of cache, if not 512KB.

Sound Card

A more expensive sound card may have more jacks (holes) on the back of it, but the real difference is in the listening. A more expensive sound card gives you better sound. If you hook up your sound card to your CD-ROM drive, you can listen to your audio CD's on your computer.

Speakers

Magnetically shielded speakers.

• HOW DO I HOOK UP MY SOUND CARD TO MY CD-ROM DRIVE SO I CAN LISTEN TO MUSIC CD'S?

There is, on every sound card, a connector for an audio cable. When you play an audio CD on your computer, the audio cable brings the sound through. Hook one end to your sound card and the other to the small connector on the back of your CD-ROM drive (it has 4 pins), and you'll have sound. Be sure your computer speakers are connected and turned on and you're in business.

• I DON'T HAVE ANY VOLUME CONTROLS ON MY SPEAKERS. IS THERE A WAY TO TURN DOWN THE SOUND?

Yes. Down at the bottom right of your screen there is a little speaker-like icon:

1. Click it once.
2. With your left mouse button, drag the slider down (that little square knob looking thing). This will lower the volume on your speakers.
3. TIP: Click the "volume control" icon twice for more options.

• I WANT TO TURN THE SOUND OFF COMPLETELY (TEMPORARILY, OF COURSE).

1. Click once on the "Volume Control" icon.
2. With your left mouse button, click on the box that says "Mute". This will turn off the sound to your speakers. Click the box once more when you are ready to turn the sound back on.

• I TRIED TO INSTALL A NEW PROGRAM ON MY COMPUTER. THE COMPUTER IS WORKING FINE, BUT THE NEW PROGRAM WON'T WORK?

When you go to purchase new software, you need to make sure that your computer meets or exceeds the system requirements to run the software, otherwise the software won't work properly. If you're not sure what your system has, do the following:

1. Click "Start"
2. Click "Control Panel"
3. Double click "System"
4. Click on the "General" tab.

It will tell you two of the most important things you need to know: how fast your computer is and how much memory you have. Something else you might need to know is whether you have enough hard drive space to completely install the program. To find this out:

1. Click "Start"
2. Click "My Computer"
3. Right click with your mouse on drive "C".
4. From the menu that comes up, select "Properties". This will tell you how much total hard drive space you have, how much of this space is taken up by programs, and how much free space you have on your hard drive.

• WHAT IS A SURGE PROTECTOR?

Another thing you might want to buy when you purchase a new computer is a surge protector. A computer (and all its stuff) needs many electrical outlets, and a surge protector provides these outlets. They call it a surge protector because it protects you if you ever have an electrical surge. Electrical surges tend to damage computer equipment.

• THE MICROSOFT® WINDOWS® XP OPERATING SYSTEM ASKED ME WHAT PROGRAM I WANTED TO OPEN A PICTURE FILE WITH. I CHOSE ONE FROM THE LIST. WHEN I DOUBLE CLICKED ON THE PICTURE TO OPEN IT, I SAW NOTHING. CAN I CHOOSE ANOTHER MICROSOFT® WINDOWS®-BASED APPLICATION TO OPEN THIS FILE TYPE WITH?

The Microsoft® Windows® XP operating system usually chooses one to open for example, music files with, without even asking you. If you should have to choose another one, and it doesn't work, don't despair. You can choose another one. Then, when you double click on the picture or file you want to open, it will open with that particular one all the time. You might need check a box (if it is there) that says, "Always use this program…".

If you want or need to change it, you can do the following:

1. Click the "Start" button
2. Click "Control Panel"
3. Go to "Folder Options"
4. Click on the "File Types" tab.
5. There will be a list of file types.
6. Go down to the one that matches the file type of the file you want to open (To do this, check out the last 3 letters of the file name you want to open (such as "myfile.???". NOTE: The "???" will of course be replaced by three other letters.) and select the one that exactly matches those letters (the letters in question will be where the question marks are. It's called a file extension.
7. Click the button that says "Change".
8. Select the new program you want to open it with and press "OK".
9. If you go to the *"Microsoft® Windows® Explorer®"* Windows®-based application and look at the file now, you will see a different icon because it now opens with a different Microsoft® Windows®-based application. Double click on the file to tell the difference.

CHAPTER 6 – THE INTERNET

If you are (or want to be) an experienced internet user, this chapter should be of help to you...

• WHAT IS THE INTERNET?

Also known as the World Wide Web. I'm sure different people would define this word differently. The internet is all those many websites out there, connected through computers and phone lines.

• I HAVE A COMPUTER AND IT IS TURNED ON. DOES THIS MEAN I AM CONNECTED TO THE INTERNET?

No. There were computers long before the internet came along. To have access to the internet, you must have a modem installed in your computer, and a program of some sort to connect you to the internet. Remember, you can make full use of a computer that does not have access to the internet.

• WHAT DOES "WWW" MEAN?

It means World Wide Web.

• WHAT DOES ".NET" MEAN?

"Net", as in ".net", means network.

• WHAT DOES ".ORG" MEAN?

It means organization.

• WHAT IS E-MAIL?

It means Electronic Mail. It means sending letters and/or pictures, etc. over the internet, like you would send a paper letter, only it's not paper.

• WHAT IS A WEBSITE/WEBPAGE?

Electronic words and pictures on the internet.

• WHAT IS A BROWSER?

It's basically, what is on your screen. The screen that your internet activities are taking place in. The Microsoft® Windows® XP operating system uses something called the *"Microsoft® Internet Explorer®"* browser.

• WHAT IS A SCREEN NAME?

A screen name is your name on the internet. You choose this when you set up your internet software.

• WHY DO I WANT AN INTERNET MEMBERSHIP?

The internet, with all of its positives and negatives, offers the ability to get information quickly, such as current events, E-mail, and you can download files that will make your equipment work better (these are called device drivers).

• IS THE INTERNET SAFE FOR ME AND/OR MY CHILDREN?

Not entirely. It all depends on how you use it. Using the internet can be very enjoyable. Some sites on the internet are very safe, fun places to visit – and some are not.

• WHAT DOES IT MEAN WHEN SOMEONE TELLS YOU TO LOG ON TO THE INTERNET?

No, you're not going to have to chop any wood. To log on to the internet means to connect to it through your modem or other connection device.

• HOW DO I LOG ON TO THE INTERNET?

You need to connect through a computer program. Most of the major internet companies send you their CD's for free in the mail.

1. Put the CD-ROM you received in the mail into your CD-ROM drive.
2. Install the software on your computer, by following the instructions that you will see in front of you.
3. Once the program is set up on your computer, you can use it to log on to the internet.

Remember, you will need to attach two things to your computer to be able to log on to the internet – a modem (which is already inside your computer when you buy it) and a telephone line attached to your modem. Logging on to the internet is literally making a phone call.

• WHY DO I NEED A PASSWORD?

Anyone who has access to your computer can log on as if they were you, if they have your password – and since sometimes you are charged by the hour, that bill can add up really fast. So, if you want to keep everyone from having access to something that is yours – you need a password.

©Ruben Gerard 1999

• WHAT SHOULD I DO IF I HAVE FORGOTTEN MY PASSWORD?

If you have forgotten the password that you use to log on to the internet, and there is no other way to get it, then you should call the company that provides you access to the internet. They can help you.

• WHY DO I NEVER SEEM TO CONNECT TO THE INTERNET AT THE SAME SPEED?

How fast you connect to the internet depends on certain factors. What kind of modem / type of connection you have, the phone number you choose when you set up your internet software, how many people are online at the same time using the same number, and the time of day. If you can't seem to log on, try again at a different time.

• OK, I'M CONNECTED. WHERE FROM HERE?

I always found the internet useful for several things. Let's say you're recording your favorite TV series and you want to know the episode titles. There are several websites that feature this.

The internet is also useful for research. If you are doing a school paper, you will find the internet helpful.

There is also one very beneficial new trend that I would like to talk about for a moment. Let's say that you want to get your college degree (whether it be your associates or bachelor's degree, or even higher for that matter), you can now attend college through your computer. This allows those who work 12 hours a day, 5 days a week to get the education they want.

• CAN I TALK ON THE PHONE AND BE LOGGED ONTO THE INTERNET AT THE SAME TIME?

No, in most cases, you can't.

• WHAT DOES DOWNLOAD MEAN?

Download means to transfer a file, picture, etc. from the internet to your computer.

• WHAT DOES UPLOAD MEAN?

Upload means you are taking a file from your computer and sending it to somewhere on the internet.

• HOW CAN I TELL HOW MUCH LONGER WILL THIS PAGE TAKE TO LOAD?

Somewhere (probably at the bottom of your screen) is a little bar. This bar starts at the left and goes all the way to the right. When it gets there, your web page will have loaded.

• WHAT DOES TRANSFER RATE MEAN?

It's how fast the information is being downloaded to your computer.

• WHAT INTERNET PROVIDER SHOULD I USE?

There is no way to say. It depends on what you want and what you like. Try out two or three of them and choose the one you like.

• WHAT IS A VIRUS?

If you contract a computer virus, it could damage your electronic equipment.

• WHAT ARE "TEMPORARY INTERNET FILES"?

When your computer goes to a website, the art (pictures, etc) on the website downloads to your computer in a particular folder. You can empty these and it might make your internet move faster. To do this:

1. Click the "Start" button
2. "Programs"
3. Click on the *"Microsoft® Internet Explorer®"* browser.
4. Click on the "Tools" menu.
5. Select "Internet Options"
6. Click "Delete Files"

• HOW DO I CLEAR MY INTERNET HISTORY?

The Microsoft® Windows® operating system keeps track of the websites you have been to for the past while. This is handy. What it translates to is when you type in the first letters of the address of the website you want to go to, different addresses come up for you with each new letter you type. Try it. You'll see what I mean.

To clear this:

1. Click the "Start" button
2. "Programs"
3. Click on the *"Microsoft® Internet Explorer®"* browser.
4. Click on the "Tools" menu.
5. Select "Internet Options"
6. Click "Clear History".

• WHAT IS A FIREWALL?

A firewall is software designed to keep people from having access to your computer.

• HOW DO I GET AN E-MAIL ADDRESS?

By the time you finish setting up your internet software (that CD you got in the mail), you will have an E-mail address. Your address is your screen name (the name you choose when you set up your account) plus "@", which means "at", then the name of the internet provider. Most internet providers will let you set up several screen names (one for personal use, one for business use, one for each child you have, etc).

There are even website networks on the internet that will let you set up a free e-mail account (which you can access from any computer). This comes in handy if all you want from the internet is to have an E-mail address.

• HOW DO I SEND AN E-MAIL?

1. Log on to the internet.
2. Click "Compose" (This will be different from program to program).
3. In the "Send To" box, type the E-mail address of the person or company you want to send your E-mail to.
4. In the subject line (a little white box near the top), type what the E-mail is about.
5. In the big white box area beneath the subject line, type your message, just as if you were typing a letter.
6. Don't forget to type your name at the bottom.
7. Click the "Send" button

You have just sent your first E-mail!

• WHAT IS AN ATTACHMENT?

An attachment is a computer file that you want sent with your E-mail. This is how families send the grandkids pictures to each other.

• I WANT TO SEND A NEW PICTURE OF THE GRANDKIDS TO MY FAMILY. HOW DO I ADD THIS TO MY E-MAIL?

1. Log on to the internet
2. Click "Compose" (or whatever button applies).
3. Type your E-mail
4. If it is a regular 3 x 5 paper photo, make sure you have used your scanner to scan it into the computer. Scan it into the file and folder of your choice (such as *"My Documents"* or *"My Pictures"*, etc.).
5. In the corner is a button that says "Attach" or "Attachments". Click this button.
6. Click "Add"
7. Find the folder with your photo.
8. When you locate it, click on it once and click "Open" … or just double click on it.
9. Click "Attach". You will then see the name of the photo in the corner.
10. When your E-mail is ready to send, click the "Send" button and your picture (and your E-mail) will be sent.

• I OPENED AND READ AN E-MAIL AND NOW I CANNOT FIND

IT AGAIN.

When you go to read E-mail. There are usually three tabs:

"New"

Mail you haven't read yet.

"Old"

Mail you have read.

"Sent"

Mail you have sent to others.

• I DOWNLOADED A PICTURE LAST NIGHT AND I WANT TO LOOK AT IT AGAIN. DO I HAVE TO DOWNLOAD IT AGAIN TO LOOK AT IT.

No. When you downloaded it, you downloaded it to your computer. So, it's on your hard drive. See the question, "I WANT TO FIND A FILE ON THE COMPUTER. HOW DO I DO THAT?" for more information.

• WHAT IS A "FILING CABINET"?

You know you can delete E-mail from your new, old and sent areas, but this is an area that you can set up to hold all the mail you have read or sent. How you set it up depends on your internet service provider. This is stored on your hard drive, so you do not have to be online to read your old mail.

• WHAT IS A LINK?

Sometimes called a "hyperlink", a link is something that you click on (with your mouse) that takes you to another web site or section of a website. A link is the words with the underline underneath them. When you put your mouse cursor over this, it changes to a little hand. Put your mouse cursor over one of these and click your left mouse button.

• WHAT IS A SEARCH ENGINE?

A search engine is a website, but unlike other websites, search engines let you find information about anything you want just by entering words into it.

Over the past several years, more and more of the popular search engines have integrated themselves into the screens that come up when you sign on to the internet.

Let's say you just logged on to the internet. Look around the screen. Near the top center of your screen, is there an area that says "Search"? Does it say, "Go" over to the right? OK. That white area. That's your search engine.

I know you want to try this out, so I'll give you a sample to use. Let's say you wanted to find out something about polar bears:

1. Click on that box (in the white part) with your mouse.
2. Type "polar bears" (without the parentheses) into that box.
3. You will see a list of links to websites and descriptions. Click on one of the links that looks good to you. It will take you to that website.

Congratulations! You just did your first internet search.

• I'VE HEARD I CAN KEEP A LIST ON MY COMPUTER OF THE WEBSITES I LIKE BEST. HOW DO I ADD SOMETHING TO MY FAVORITE PLACES LIST?

1. Go to the website you want to add to the list by typing the address into your browser and pressing "Go".
2. Click the "Minimize/Maximize" button in the top right corner of your screen until you see a heart of some kind.
3. There is an item (which is different from internet service to internet service, but usually the left most item right next to the minimize button). Click this item with your mouse.

• HOW DO I REMOVE SOMETHING FROM MY FAVORITE PLACES LIST?

1. Look at your favorite sites list.
2. Click on the one you want to delete (one time, with your left mouse button).
3. Press the "Delete" button on your keyboard (just to the right of your "Enter" or "Return" key). You can also click once on the item with your left mouse button, then right click on the item with your mouse and select "Delete".
4. When it says, "Are you sure…" click "Yes".

• WHAT IS RELOADING?

Sometimes when you go to a website, you have occasion to want the page to reload. What this means is simply having the website come up on your computer screen again, just like the first time you went there.

• WHAT ARE COOKIES?

Cookies are files from the websites you visit that help them work properly. Note that some cookies will tell others (over the internet) information about you – such as your name.

• I NEED TO RE-WORK MY SETTINGS FOR ACCEPTING COOKIES. HOW DO I DO THIS?

1. Click on the "Start" button

2. Click "Programs"

3. Click on the *Microsoft® Internet® Explorer®* browser.

4. Click on the "Tools" menu.

5. Click "Internet Options"

6. Click the "Privacy" tab

7. You will see a slider. Click and hold your left mouse button on this.

8. Drag the slider up or down to the desired level.

9. Click "Apply"

You're done.

• I ADJUSTED MY SETTINGS TO ACCEPT COOKIES, AND I'M STILL HAVING TROUBLE. WHAT ELSE CAN I DO?

Try going to the site by first going to the *"Microsoft® Internet Explorer®"* browser:

1. Click on the "Start" button.

2. Click on "Programs"

3. Click on the *"Microsoft® Internet Explorer®"* browser.

4. Enter the address (url) of the website you want to go to.

This should eliminate your problem. Every time you need to visit a certain site that has trouble with cookies and nothing else works, do the above.

• CAN I BUILD MY OWN WEBSITE?

Absolutely you can.

• WHAT IS A SERVER?

It is a place on the internet that can be free or someplace you have to pay for the opportunity to store your website files there.

• WHAT DO I NEED TO BUILD MY OWN WEBSITE?

You need a few things...

1. An internet connection
2. You need a program to build it with. Go to your nearest computer software store and ask for website building software.
3. Somewhere on the internet to store your page(s) and files (a server).

• WHAT DOES THAT LONG INTERNET ADDRESS MEAN?

Here is a sample address:

Www.soandsoserver.com/myscreenname/home/mypage.abc

Here is a breakdown – one step at a time...

"Www.soandsoserver.com" is the group of people that host your website. The name will obviously be different.

"Myscreenname" is the screen name you came up with for yourself when you signed up. Just so there is no confusion, you obviously wouldn't type "myscreenname", you would type whatever you decided it was going to be. I'm just using that as an example.

"Home" or some other possible name might be the folder where your main page is stored. You can choose whatever name you want, of course. The point being, you have a certain number of megabytes of web space to store the folders, pictures and files of your website. Other pages might be stored in folders called "page2", "page3", or whatever. For example, if you wanted to go directly to the page you have named "page2", the address that you would type would be: www.soandsoserver. com/myscreenname/page2/mypage.abc

(without any periods after it).

"Mypage.abc" is the name of the file that is, literally, your page.

• WHAT IS AN INTERNET CAMERA?

Internet cameras were intended to help you communicate with people. If you have a lot of family, but they're spread out over the 50 states, then this invention is for you. When you hook the internet camera up to your computer, you can see the person you are communicating with – face to face.

• OKAY, I'M DONE. NOW HOW DO I LOG OFF?

1. In the upper left hand corner, click "File".
2. Click "Exit" or "Close".

• IS THERE ANOTHER WAY TO LOG OFF THE INTERNET?

Yes, with your left mouse button, go up to the very tiptop left corner and double click.

• I HAVE A GAME THAT CAN ONLY BE PLAYED OVER THE INTERNET.

This is called online gaming. You pay for a monthly membership fee and you can play the game alone or against other people over the internet.

• THAT'S NICE. CAN I PLAY THIS GAME ON MY OWN COMPUTER WHEN I'M LOGGED OFF OF THE INTERNET?

No.

• WHAT IS AN INSTANT MESSAGE?

It's a way of communicating that is faster than E-mail. You can send an instant message to anyone who is online when you are.

• HOW DO I SEND AN INSTANT MESSAGE?

This differs depending on your internet service.

1. There is a button or selection in your menus for this.
2. Press it and a little box comes up.
3. Enter the screen name of the person you are sending the instant message to in the small box (that is probably at the very top). This person usually uses the same internet provider that you do.
4. In the larger box, type what you want to say to this person.
5. When you are done typing, press "Send" (or whatever button applies to your internet service).
6. The other person receives this message and can send you one back to you after he or she reads it.

• CAN I SEND INSTANT MESSAGES BACK AND FORTH WITH MORE THAN ONE PERSON AT THE SAME TIME?

Yes. Remember that button you pressed to bring up the instant message box in the first place? Press it again.

• IS THERE A WAY TO TELL IF A PARTICULAR SOMEONE IS ONLINE SO I CAN SEND THEM AN INSTANT MESSAGE?

Yes, with the America Online® service Buddy List® feature, you can make a list of people that you routinely talk to. When they log on, you will see their screen name appear.

• I WANT TO USE THE AMERICA ONLINE® SERVICE BUDDY LIST® FEATURE, BUT I DON'T HAVE ANY PEOPLE ADDED YET. HOW DO I ADD PEOPLE TO THIS LIST?

Okay…

1. Click the "Setup" button.
2. There are 3 headings listed for you to list people under.
3. Double click on the heading that you want to add people under.
4. Click "Add".
5. Enter the person's screen name into the box.
6. Click "Save".

Repeat if you want to add more people.

• HOW DO I DELETE SOMEONE FROM THE AMERICA ONLINE® SERVICE BUDDY LIST® FEATURE?

1. Click the "Setup" button.
2. There are 3 headings.
3. Double click on the heading that you want to delete people under.
4. Highlight the person's screen name with your mouse.
5. Click "Delete"
5. You will be asked, "Are you sure…?".
6. Select "Yes".

• OKAY. I'M USING THE AMERICA ONLINE® SERVICE BUDDY LIST® FEATURE AND I SEE THAT THEY ARE ONLINE, HOW DO I SEND HIM OR HER AN INSTANT MESSAGE?

1. Double click on their screen name and a box comes up.
2. Now you can type and send them an instant message as described above.

• I'M DOING FINANCIAL TRANSACTIONS ONLINE. IT SAYS I NEED A BANK ACCOUNT. I DON'T WANT TO GIVE OUT THAT INFORMATION ONLINE.

The best thing to do here is to go down to your bank and open a bank account that is just for online transactions. That way you won't have to give out your main account number.

CHAPTER 7 – WORKING WITH THE MICROSOFT® WORD® WINDOWS®-BASED APPLICATION

The Microsoft® Word® Windows®-based application is one of the most popular Microsoft® Windows®-based applications of its kind on the market today. Since there are many people out there who are going to want to do word processing on their new (or even their old) computer, this chapter should be beneficial.

• WHAT IS THE MICROSOFT® WORD® WINDOWS®-BASED APPLICATION?

It is a for word processing.

• WHAT IS HIGHLIGHTING?

That's dragging the mouse over your text (to "mark" it).

• HOW DO I START A NEW DOCUMENT?

If the new document is not already started for you:

1. Click on "File"
2. Click on "New"
3. If you need to, click your mouse in the top left hand corner of the page to get started.

OR

1. Click on the little white piece of paper in the top left corner under the word "File".

• HOW DO I OPEN A FILE?

1. "File"
2. "Open"
3. Select the file you want to open and click "Open".

OR

1. Click on the folder at the top of the screen.

• HOW DO I SAVE A FILE?

1. Click on "File"
2. Click "Save As"
3. If you want to change the name of the document, you can type a new one.
4. Click "Save"

OR

1. Click on the little disk at the top of the screen.

HELPFUL TIP — Try saving a file this way:

1. Press "Alt" and "F" at the same time or "Alt" and the down arrow. (This activates the "File" menu).
2. Use the down arrow to select "Save As".
3. If you are just saving the document again as you did before (meaning with the same name), press "Enter" twice.

4. See... you never even had to touch your mouse.

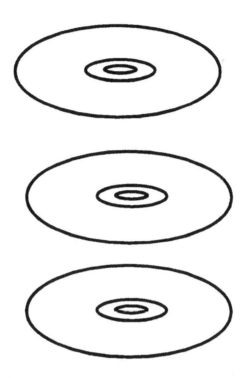

• I SEE THAT THERE IS TWO CHOICES TO SAVE A DOCUMENT – "SAVE" OR "SAVE AS". WHAT'S THE DIFFERENCE?

If you use "Save", you have most likely kept the same name as the file had before. "Save As" is usually for saving the file under a different name.

• HOW DO I PRINT A DOCUMENT?

1. Click on "File"

2. Click "Print"

3. Click "OK"

• I JUST TYPED A LETTER AND I WANT TO RUN A "SPELLING AND GRAMMAR CHECK". HOW DO I DO THAT?

1. "Tools"
2. "Spelling and Grammar"

If you don't want it to check the grammar but you want it to check the spelling, un-check the box that says, "check grammar".

• I SEE RED AND GREEN LINES IN MY DOCUMENT.

The "Spelling and Grammar" checker has a dictionary of it's own. The red and green lines you see in your document are telling you that, according to the internal dictionary, you have spelling and/or grammar errors. The red lines are spelling errors, the green lines are grammar.

• IF I PRINT OUT MY DOCUMENT AS IT IS, WILL THESE RED AND GREEN LINES SHOW UP ON MY PRINTOUT?

No, they will not.

• I WANT TO USE A DIFFERENT FONT. HOW DO I CHANGE IT?

At the top of the screen, just to the right of center there is a white box. Click your mouse on the little arrow going down and click on the font you want. From that point forward, you will be typing in that font.

• I WANT TO CHANGE THE FONT SIZE (HOW BIG IT IS). HOW DO I DO THAT?

Just to the right of where it says "10", up at the top of the screen, there is a down arrow with many numbers. The number you select represents the size of your text. The larger the number, the bigger the text.

• I WANT TO HAVE MY TEXT IN BOLDFACE. HOW DO I DO THAT?

One of two ways:

1. Click the "B" at the top of your screen and start typing.

OR

1. Drag your mouse over the text you want to boldface, and press the "B" at the top of your screen.

• I WANT TO HAVE MY TEXT IN ITALICS. HOW DO I DO THAT?

One of two ways:

1. Click the "I" at the top of your screen and start typing.

OR

1. Drag your mouse over the text you want to put in italics, and press the "I" at the top of your screen.

• I WANT TO UNDERLINE MY TEXT OR UNDERLINE SOME TEXT THAT I DIDN'T UNDERLINE BEFORE. HOW DO I DO THAT?

One of two ways:

1. Click the "U" at the top of your screen and start typing.

OR

1. Drag your mouse over the text you want to underline, and press the "U" at the top of your screen.

• I WANT TO CHANGE THE COLOR OF MY TEXT. HOW DO I DO THAT?

1. Click on the little down arrow next to the "A" with the little colored bar under it and select a different color

OR

1. Highlight text with your mouse and select a different color

OR

1. Click on "Format"
2. Click on "Font"
3. Click on the down arrow next to where it says "Automatic", select a different color, and click "OK".

• WHAT IS SCROLLING AND HOW DO I DO IT?

Scrolling is moving in your document. You can do this in one of several ways. Assuming we are starting at the top left hand corner, to scroll down you can:

1. Press "Page Up" or "Page Down"

2. Use the up or down arrow keys.

3. Look at the gray strip at the right side of your screen. You can drag that little box/block up or down to see other pages or you can click with your mouse above or below that box/block to move to other pages.

• I HAVE A VERY LARGE DOCUMENT THAT I AM EDITING AND I SCROLLED DOWN REALLY FAR, THEN I PRESSED ONE OF MY ARROW KEYS AND THE CURSOR WENT RIGHT BACK TO WHERE IT WAS BEFORE.

Make sure to click on that page with your mouse when you are where you want to be. If you, for instance, press the down arrow key before you click on the new location with your mouse, you will go back to where you were before and have to find that page you just took time finding all over again.

• I HAD MY FONT SIZE (SIZE OF TEXT) SET TO A LARGER SIZE AND NOW ITS SMALL AGAIN.

In the Microsoft® Word® Windows®-based application, if you go up to an earlier point in your document your font size might change. Ask yourself, "What did I have that number set at when I was typing this paragraph?". If you want to make sure that all your text is the same size to begin with, the best way is to set that number at the beginning of typing your document, before you start typing. That way, you'll have no problems.

• THERE IS A PARTICULAR WORD OR PHRASE I WANT TO FIND OR GO TO IN MY DOCUMENT. WHAT IF I WANT TO GO TO A SPECIFIC PAGE? HOW DO I DO THAT?

1. Click "Edit"
2. Click "Find"
3. You will see three tabs –

1. "Find" (you use this one to find a word you are looking for in the document). Let's say you were looking for all the different times you typed the word *for*. If you enter the word *for* into the box and click "Find Next", you will see the first time you typed the word, click "Find Next" again to see the second time you used the word, and so on.

2. "Replace" (you will use this tab not only to find words, but also to replace those same words with others. For example, let's say you wanted to replace all of the times you used the word "for" with the word "but". In the "Find What" box you'd type "for", and in the "Replace With" box you'd type "but". When you got to each word you want to replace, you'd hit the "Replace" button.

3. "Go To" will let you go to a specific area of your document. For example, if you were on page 1 of your document and you wanted to go to page 20, you would enter "20" in the box and press "Next".

• THE MENUS HAVE BEEN CUT OFF!

Some versions of the Microsoft® Word® Windows®-based application will learn. They will cut off the options you don't use and the other ones will be hidden. If you want to see the whole thing again, just hold your mouse arrow over the arrows pointing down at the bottom of the menu and you will see the whole thing.

• WHAT ARE ALL THOSE BUTTONS AT THE TOP?

Each one of those pictures does something if you click on it. You can put your mouse arrow on each one to find out what it does.

Here is what some of the "pictures" mean:

The little piece of paper lets you open a "New" document.

The file folder lets you "Open" a document already on your computer.

The little disk lets you "Save" a document.

The paper and magnifying glass is the "Print Preview" button.

The two pieces of paper is the "Copy" command.

The little clipboard is the "Paste" command (as in copy and paste).

The little half round arrow will "Undo" the last typing you did.

The little globe and chain link is for insert an internet link.

The little 100% is the "Zoom" command.

The little white area at the top (about half way over) is the font list. You can change this if you want.

The area that says "10" is the font size

"B" means "Boldface".

"I" means "Italics".

"U" means "Underline".

The one with all the little lines is "Align Left" and the one next to it is center.

The one with the three vertical dots is the "Bullets" command.

The "A" with the underline is to change the "Font Color.

• WHEN YOU ARE SAVING A DOCUMENT AND IT SAYS "SAVE IN THIS OR THAT FORMAT", WHAT DOES IT MEAN?

Having a file in a particular format (not to be confused with formatting a floppy disk) is saving your file in a particular way (i.e. as a particular file type). How easily the Microsoft® Word® Windows-based application reads it or whether you can read it in other Microsoft® Windows®-based applications depends on what format you save it in. If you save it in the default format it will be more easily read. There are several options listed (when you go to save your files, you will see this). Just as an example, (this is not the default format, but) if you were to save something in a "text only" format, almost any Microsoft® Windows®-based word processing application would read it.

• I HAVE TWO DOCUMENTS I WANT TO MAKE INTO ONE.

1. Put your cursor at the point where you want to insert the other file.
2. Click on the "Insert" menu at the top of the screen and scroll down until you see the word "File".
3. Type the name of the file you want to insert in the white box next to the word "Range".
4. Click the "Insert" button off to the right.
5. Save your new file.

• IS THERE ANY WAY TO AVOID LOSING A DOCUMENT DUE TO A POWER OUTAGE?

Yes, there are large power batteries that you can hook up to your computer that activate quickly enough when you lose power in your home that your computer stays on, so you can save your document. The best advice is to save your documents after you have made significant progress, so they will be stored on your hard drive when you need them.

• I HAVE TWO DIFFERENT VERSIONS OF THE FILE I'M WORKING ON. HOW DO I TELL WHICH ONE IS THE ONE I WORKED ON LAST?

Let's say you have one version of the file on your hard drive and the other on you USB drive. What you need to do to find out the last one you saved is this; let's say you want to check out the one on drive C:

1. Go into the *"Microsoft® Windows® Explorer®"* Windows®-based application. You forgot how? No problem...
2. Click "Start"
3. Click "Programs"
4. "Accessories"
5. Click on the *"Microsoft® Windows® Explorer®"* Windows®-based application.

You don't see drive "C"? Again, no problem...

6. Click on "My Computer" on the left side of your screen.
7. You will see your list of drives.
8. Click on Drive "C"
9. Locate your file
10. Right click on your file with your mouse.
11. Select "Properties"
12. It will tell you the day and time that your file was last saved.
13. Now do the same for the file on your USB key.

• WHAT IS A TABLE?

Let's say you need to sort some information for your accounting course. A table is a way to make your information look nice. It sorts all your information in a very asthetically pleasing manner.

• HOW DO I CREATE A TABLE?

1. Click on the "Table" menu at the top of the screen.
2. Select a point at the left side of the screen to begin your table.

3. Hold down your left mouse button and move your mouse from the left side of your screen to your right.

4. Now let go of your left mouse button. You have drawn the outside edges of your table.

5. You want to split it up into sections, so...

6. Physically draw the inside of the table to with your mouse in the same way that you created the outside edges. Hold down your left mouse button and move it from the left side to the right side and let go of your mouse button.

7. Voila! You have your table!

8. Now click with your mouse on the section of your table that you want to put information into and type the information. It's that easy.

• WHAT DOES "PRINT PREVIEW" DO?

It lets you see what the current page is going to look like after it gets printed. The up side is – you get to see what it looks like before it gets printed, so you can make changes if you want.

• WHERE IS "PRINT PREVIEW?"

When you already have a document open...

1. Click on the "File" menu.
2. Click "Print Preview".

• I'M TYPING MY RESUME AND I WANT TO INCLUDE MY PICTURE. CAN I DO THAT IN THE MICROSOFT® WORD® WINDOWS®-BASED APPLICATION?

Of course you can.

Remember the menus at the top of the screen?

1. Click on the "Insert" menu.

2. Select "Picture"

3. Select "From File"

4. Now you have to search for the file. If it's inside a folder, remember to double click on the folder to open it.

5. Select the file and click "Insert".

That's all there is to it.

CHAPTER 8 – MAKING FILES SMALLER

What with downloading all those huge drivers and scanning all those graphics files, this might come in handy.

• WHAT IS A COMPRESSED FILE (OR FOLDER)?

A compressed file (or folder) is a file (or folder full of files) that has been shrunk to make it fit on a disk that it might be too large to fit on otherwise.

• DO I NEED A SEPARATE MICROSOFT® WINDOWS®-BASED APPLICATION OR CAN I MAKE A COMPRESSED FILE (OR FOLDER) IN THE MICROSOFT® WINDOWS® XP OPERATING SYSTEM?

There are several good Microsoft® Windows®-based applications out there to do this, but you can also compress files in the Microsoft® Windows® XP operating system:

• HOW DO I CREATE A COMPRESSED FOLDER FULL OF FILES? (I DON'T HAVE A FOLDER FULL OF FILES YET).

If you don't already have a folder full of files to compress, you need to create a folder and drag and drop files to it. So let's review…

1. Go to the *"Microsoft® Windows® Explorer®"* Windows®-based application and click on "My Computer".
2. Double click with your mouse on the drive on which you want to create the folder (such as drive "C").

3. Right click on your Microsoft® Windows® operating system desktop and move your mouse over to the word "New"

4. Move your mouse over to the word "Folder", and click.

5. A generic folder will come up waiting for you to type a name. Type a name for the folder and press "Enter".

6. We will name this folder "compressed1".

7. Double click on the new folder to open it.

8. Copy (drag and drop) files to the new folder.

9. Right click on the folder that has the file in it.

10. Move your mouse over to the "Send to" option. On the list, it will say "Compressed folder".

11. Click on the compressed folder with your left mouse button.

12. The Microsoft® Windows® XP operating system will create a separate, duplicate (but compressed) folder with the same name and files as the one you created in step 5.

13. To look to see if your files have been compressed, double click on the folder with the zipper on it. You will see the folder with your files in it.

• WHAT IF I ALREADY HAVE A FOLDER OF FILES THAT I WANT TO COMPRESS?

1. Right click on the folder and select the "Send To" menu option.

2. Then select "Compressed Folder" and click on it with your left mouse button.

• HOW DO I DECOMPRESS A COMPRESSED FOLDER?

1. Go to the *"Microsoft® Windows® Explorer®"* Windows®-based application.

2. Find the folder and click on it.

3. To your left you will see "Extract All Files".

4. Click on "Extract all files".

5. The Microsoft® Windows® XP operating system will bring up the "Compressed Folders Extraction Wizard". This will lead you through the steps to extract (decompress) all of your files.

• THERE'S A FILE THAT I FORGOT TO ADD TO THAT COMPRESSED FOLDER!

No problem. Just drag and drop it to the folder you want it in. NOTE: This works best when you can see your folder and the file you want to add at the same time.

• IS THERE A QUICKER WAY TO COMPRESS A SINGLE FILE TO A NEW COMPRESSED FOLDER?

Yes. If you right click on the file you want to compress and select the Send To option, it will create a compressed folder with the compressed file in it. This folder will have the name of that file.

CHAPTER 9 – BURNING CD'S OR DVD'S

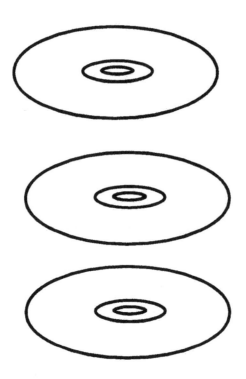

• WHAT IS A CD / DVD BURNER AND WHAT DOES "BURNING" A CD OR A DVD, MEAN?

A CD / DVD "burner" is a drive that can store data, pictures, music, etc. to a CD-ROM or DVD-ROM disk. The different types of disks are mentioned in this chapter.

• WHAT IS WRITING TO A DISC / DISK?

Writing to a disc (of any kind) means saving information to it.
This doesn't mean taking your pen and writing on a label.

• WHAT IS CD-R?

CD-R means Compact Disc Recordable. It means you can put information on it only once and can't erase it. You can add info to it until it is full, but you can't erase it. A CD-R usually holds 74 or 80 minutes of music and more than megabytes (MB) of data. These are primarily used in a CD burner.

• WHAT IS CD-RW?

CD-RW means Compact Disc Re-Writable. You can format this disc much like you format a floppy disk and you can write information to it repeatedly, and erase that information repeatedly. These are primarily used in a CD burner. These discs hold the same amount of information as a CD-R.

• WHAT IS DVD-R?

When you get into DVD media, you have several different kinds of discs and each of the different brands of drives you might buy support different disks, meaning some may work in one drive and some won't work in that one, but they'll work in a different one.

DVD-R (Digital Video Disc [minus] Recordable) is a DVD that you can write to once and can't erase that information. DVD-R's hold 4.7 gigabytes of information.

• WHAT IS A DVD+R?

You can write to it once. It holds the same amount of information as a DVD-R.

• WHAT IS A DVD-RW?

This is like a CD-RW, but it holds much more information: 4.7 gigabytes of information to be exact. You can erase and re-record on this as much as you like.

DVD-R's, DVD+R's, and DVD-RW's can also be used in DVD recorders (sort of like a VCR, but with a DVD in it). They hold 2 hours, 4 hours, or 6 hours of video depending on the speed you record at — just like videotape.

• CAN I BURN CD'S USING THE MICROSOFT® WINDOWS® XP OPERATING SYSTEM?

Yes. There is a Microsoft® Windows®-based application similar to one called the Roxio® EZ CD Creator® — in the Microsoft® Windows® XP operating system.

• HOW DO I BURN FILES TO A CD OR DVD IN THE MICROSOFT® WINDOWS® XP OPERATING SYSTEM?

1. Put a blank CD-ROM or DVD into your CD or DVD burner.
2. Find out which drive on your computer is your CD burner or DVD burner.
3. Go to the *"Microsoft® Windows® Explorer®"* Windows®-based application and select the files you want to copy to the CD or DVD by clicking on them with your mouse. To select more than one file, hold down "CTRL (Control)" while clicking on the files.
4. From the "Edit" menu, select "Copy",
5. Click on Drive "E" (or whichever drive your CD burner is).
6. From the "Edit" menu, select "Paste".
7. It will show you a list of files waiting to be written to the CD.

8. Under the "File" menu, select "Write These Files To CD".

9. The CD writing wizard will come up and ask you what you want to name the CD. Choose a name and click "Next".

10. It will then write the files to the CD and "finalize" the disk.

11. It will then give you the option to write them to another CD or not. If you don't want to, click "Finish" and you're done.

• WHAT IS FINALIZING A DISK?

Unless you want to read a disk you are burning in the drive in which you created it, it cannot be read in a regular CD-ROM drive until you "finalize" it. Don't worry, the software you are using will take care of it, if you follow the instructions, but you must finalize — or "close" — the disk if you want it to be read in any other kind of computer drive.

• WAIT! THERE'S A FILE ON THERE I DON'T WANT WRITTEN TO THE CD.

That's okay.

1. Click on the file with your mouse.

2. Press the "Delete" key on your keyboard.

3. When it asks you if you are sure you want to delete this item, select "Yes".

CHAPTER 10 – DIGITAL CAMERAS AND WORKING WITH PICTURES

Photos are a part of everyone's life. Just as we have to change with the times, so does the camera...

• WHAT IS A DIGITAL CAMERA?

A digital camera lets you take pictures and transfer them to your computer. First, let's talk about quality. Digital cameras are measured in what they call megapixels.

Digital cameras have become very affordable these days. They have increased in quality from 2.1 megapixels, to 3.0 and 3.1 megapixels, to 4.0 and most currently, to 5.0 megapixels.

• HOW DO I GET THE PICTURES FROM MY DIGITAL CAMERA TO MY COMPUTER?

Most digital cameras come with a small hole in them. The hole has a cable attached to it. The cable goes from the hole in the camera to a USB port

1. Turn on your computer
2. Go into the "Control Panel".
3. Go to "Scanners and Cameras"
4. Turn your digital camera on.
5. The Microsoft® Windows® XP operating system will recognize the camera.
6. Double click on the picture of the camera.
7. If it asks you which program you want to use, double click on one of the choices listed or click on a choice and click "OK".
8. The "Scanner and Camera Wizard" have come up. Click Next.
9. Select all the pictures you want to transfer to your computer.

10. Select a name for the file and a folder for them to go into. Remember to click the down arrow to the right of what is listed to see the dropdown menu.

11. Select "Next"

12. Next, you will see a list of options, if you just want to go look at your pictures, highlight "Nothing"… and click "Next".

13. Click "Finish" or click on the link with the name of the folder and it will take you to the folder you saved your pictures to and show you your pictures.

NOTE: Many times, the Microsoft® Windows® XP operating system will, by default, save your pictures to the *My Pictures* folder, located within the "*My Documents*" folder.

• I CAN MAKE MOVIES WITH MY DIGITAL CAMERA?

Yes, some digital cameras have a setting where you can make very short video clips.

• WHICH DO YOU RECOMMEND FOR MY DIGITAL CAMERA – DIGITAL ZOOM OR OPTICAL ZOOM?

Optical zoom.

• WHAT IS A MEMORY CARD?

It's a place to store your photos when you don't have enough space in your digital camera's internal memory.

• I NEED TO BUY A MEMORY CARD FOR MY DIGITAL CAMERA. HOW DO I DETERMINE WHAT SIZE TO BUY?

They come in the following sizes: 32 megabytes, 64 megabytes, 128 megabytes, and 256 megabytes. What size you buy depends on two things – how many pictures your are going to take and what quality setting your camera is set to. If you have your camera set to a higher quality setting, your camera (or your memory card) is going to hold less pictures.

• WHAT'S THE PURPOSE OF A MEMORY CARD READER?

A memory card reader is used in case you don't have access to your camera, but you want to transfer pictures to your computer. You hook it up to your computer and…voila!

• DO ALL DIGITAL CAMERAS USE THE SAME TYPE OF MEMORY CARD?

No. You must check your digital camera's manual to determine what kind of memory card your digital camera will accept, otherwise the card might not fit.

• WHO AM I GOING TO GET TO TAKE PICTURES OF ME?

Most digital cameras these days come with a timer on them.

• WHAT OTHER ACCESSORIES CAN I BUY THAT WILL BE USEFUL?

Well, if you live by yourself and you don't have anyone to take pictures of you, you might want to try buying a tripod.

• I NEED TO BUY BATTERIES FOR MY DIGITAL CAMERA. WHERE CAN I BUY THEM?

Any electronics store should have them.

• I HAVE SOME OLD PHOTOS THAT ARE FALLING APART. CAN COMPUTERS HELP ME TO PRESERVE THEM?

Absolutely. You can use your scanner to preserve these photos.
If you want to preserve them exactly as they are, all you have to do is put them in your scanner and save them as files. It is possible to clean them up – make dark photos brighter, faded ones look like new, but you have to buy computer software capable of photo editing. Call up your local software store and ask.

• WHAT ARE THUMBNAILS?

Thumbnails are just smaller versions of photos. Something like you might see in the *Microsoft Windows® Explorer®* Windows®-based application.

CHAPTER 11 — TROUBLESHOOTING TOPICS

When something goes wrong with a computer, it's usually something simple. Something may be unplugged, not connected properly. It could be something more serious…but it isn't usually.

The most common problem that you might run into is the following one:

• I TURNED BY COMPUTER ON, BUT I DON'T SEE THE MICROSOFT® WINDOWS® OPERATING SYSTEM.

Do you have a floppy drive on your computer?

Check it.

Is there a disk in it?

If the answer is "yes" – take out the disk and press a key on your keyboard.

• MY COMPUTER IS FROZEN AND I CAN'T TURN IT OFF! WHAT DO I DO NOW?

You can do one of two or three things. You can:

1. Unplug it
2. If you have a surge protector, turn the master switch (the big one on the top) off and then turn it on again.
3. Some computers will turn your computer off if you press hard on the power switch and hold it down. This will not hurt your computer, but it will give you the same result as shutting down your computer improperly.
4. Try pressing "Control Alt Delete" (all three keys at the same time). Highlight the frozen program with your mouse, then press "End Task" and your computer will start working again (See the question, "WHAT ARE ALL THESE EXTRA KEYS ON MY KEYBOARD?" for more info.

• I WANT TO SEE WHAT'S INSIDE A FOLDER. I CLICKED ON IT, BUT IT DIDN'T OPEN. WHAT'S WRONG?

Double click on the folder with your left mouse button.

• I THINK I HAVE A COMPUTER VIRUS. WHAT DO I DO?

Go to your nearest office supply store (one that carries computer software). Ask an associate to find you a program to get rid of computer viruses.

• MY COMPUTER IS RUNNING FINE, BUT THERE'S NOTHING ON MY COMPUTER SCREEN.

One of a couple of things is happening:

1. The cable that goes from the back of your monitor to your computer is not plugged in all the way. In this case, turn your computer around so you can see the back and check to see that your cable(s) are plugged in properly.
2. The computer could be in "sleep mode". What this means is after the computer has been left untouched (the mouse has not been moved) for a certain amount of time, the computer screen shuts down. In this case, you just have to move the mouse.

• THE PICTURE ON MY COMPUTER MONITOR IS OFF TO ONE SIDE.

The controls on your monitor need to be adjusted. On the front of most monitors, there are controls for brightness, contrast, and vertical and horizontal controls. Check the manual that came with your monitor to adjust the position of your picture until it is where you want it.

- **I MOVED MY CURSOR AND I CLICKED IT IN THE LITTLE BOX TO BE ABLE TO TYPE SOMETHING. THE CURSOR IS BLINKING, BUT I PRESS THE KEYBOARD KEYS AND THE LETTERS/NUMBERS DON'T SHOW UP ON THE MONITOR.**

Move your mouse over to it and try clicking in the dialog box again and then try typing again. It should work.

Also, remember to make sure that your number lock is on.

- **HOW DO I MAKE A BOOT DISK/START-UP DISK (WITHIN THE MICROSOFT® WINDOWS® XP OPERATING SYSTEM)?**

A boot disk is something that is usually used for an emergency of some kind where your computer will not come up normally. Take note that you usually need a CD-ROM driver on this disk also to make that work (because the Microsoft® Windows® XP operating system is on a CD, remember?). Hopefully, you are making this disk before there is a problem.

1. Go to "My Computer"
2. Right click on drive "A":
3. Select "Format"
4. Be sure to check the box that says, "Create An MS-DOS® Startup Disk". You might want to uncheck this after you are done.
5. Press "Start"

Now, if you just want to boot up to drive "A": and don't need to boot up to your CD-ROM with this floppy disk, you can skip the next steps:

1. Go to the internet, type the words CD-ROM drivers into your search engine, and download a generic one.

2. Download the CD-ROM driver to the same boot disk you just made.

3. Look at the files on this disk using "My Computer" or the *"Microsoft® Windows® Explorer®"* Windows®-based application.

4. Double click on the file. The file you downloaded is compressed. It will de-compress.

You now have a boot disk (start-up disk) to use in emergencies.

• I JUST USED MY BOOT DISK TO BOOT MY COMPUTER AND IT BOOTED UP TO A MICROSOFT® MS-DOS® OPERATING SYSTEM COMMAND PROMPT. HOW DO I CHANGE TO DIFFERENT DRIVES ON MY COMPUTER?

To change to drive "A", enter the command: "A:" and press "Enter".

To change to drive "B", enter the command: "B:" and press "Enter".

To change to drive "C", enter the command: "C:" and press "Enter".

To change to drive "D", enter the command: "D:" and press "Enter".

To change to drive "E", enter the command: "E:" and press "Enter".

To change to drive "M", enter the command: "M:" and press "Enter".

If you are in a subdirectory such as "C:/miscellaneous/other" and you want to change to the root directory (such as "A" or "C"): you need to type one more thing after the letters above. Type the "/" key (for example "A:/" or "C:/") and it will take you to the root directory of that drive.

• MY COMPUTER HAS REALLY SLOWED DOWN. WHAT HAPPENED?

When you use your computer programs a lot, or even if you don't and you install many programs, all the files that were once installed that were "together" (meaning back to back) before, have now split in 10 different directions. This is what's called fragmentation. To put it simply, your hard drive is fragmented. To make your computer faster again, you need to defragment your hard drive.

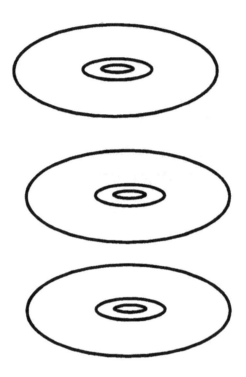

• HOW DO I DEFRAGMENT MY HARD DRIVE?

There is a Microsoft® Windows®-based application for this included on your computer system.

1. Click "Start"
2. Go to "Programs"
3. Go to "Accessories"
4. Go to "System Tools"
5. Click on "Disk Defragmenter"

6. Drive "C" is already going to be selected, so just hit the button that says, "Defragment".

• WHAT DO I DO IF MY COMPUTER IS STILL RUNNING SLOWLY AFTER I HAVE DEFRAGMENTED THE HARD DRIVE?

Try removing unused programs from your hard drive.

• MY KEYBOARD IS DUSTY (INSIDE AND OUT)! DO YOU HAVE ANY SUGGESTIONS TO CLEAN IT?

To clean the outside, any dusting spray will do.

To clean the inside, office supply stores sell cans of compressed air. You just spray this (in this case, between the keys) and it will get all the dust out.

• HOW DO I CLEAN MY CD-ROM (THE CD-ROM DISK)?

From time to time CD-ROM's get fingerprints and/or dust on them. The most common way to clean them is to breathe (exhale) on them and then wipe.

• WHAT'S THE BEST WAY TO KEEP DUST FROM GETTING INSIDE MY COMPUTER?

Everything comes out the back of the computer right? For those slots that aren't being used, there is a gray "cover" that screws on. You screw these on inside the computer. If you have open slots, you need to put these covers on. If you have none, you can ask at your local electronics store. Covering these slots will keep out dust.

• I CAN'T TURN MY COMPUTER ON AND I NEED TO GET THE CD-ROM OUT OF THE DRIVE.

Here's what you do:

1. Get down on your knees and take a close look at the front of your CD-ROM drive. One of those holes is used in case of this particular situation.
2. Go find a paper clip and pull on it so that the clip is now straight for about an inch to an inch and a half.
3. Push the end of the paper clip into the hole.
4. Your CD-ROM will pop out and you can retrieve it.

• I JUST TRIED TO INSTALL A NEW VIDEO CARD INTO MY COMPUTER AND NOW MY COMPUTER DOESN'T WORK.

Your video card probably isn't seated properly in the slot you put it into.

1. Take it back out and put it back in again. Put one hard on each part of the card as you're pushing it in. It should give you a little resistance as it's going in, but it should go in.

• I WENT TO HOOK UP MY MOUSE AND THE CONNECTOR ON THE MOUSE AND THE CONNECTOR ON THE BACK OF THE COMPUTER DON'T MATCH!

You can go to your local computer supply store and buy a mouse adapter.

• I JUST TRIED TO READ THE INFORMATION ON MY FLOPPY DISK AND THE COMPUTER WON'T READ IT!

From time to time, when you try to put a disk in it may not be read for a variety of reasons, including:

1. It may be a bad (damaged) floppy.

2. Your floppy drive may be damaged or need cleaning. You can clean it by going out and buying a cleaning disk, available at an office supply store.

3. The most common occurrence is that your floppy may have been exposed to a magnetic field of some kind. This tends to erase information on floppy disks.

• I JUST BOUGHT A NEW SCANNER AND I CAN'T GET IT TO POWER UP. IS IT BROKEN?

Electronics today usually come with a detached power cord. When you pulled it out of the box, some nice person probably already plugged it in for you. When it's all set up, and if it still doesn't work, check to make sure the power cord is very firmly plugged in (at both ends). This should solve the problem.

• HOW COME MY SCANNER DOESN'T SCAN IN COLOR?

Don't worry, your scanner isn't broken. Make sure that the application that you are using is set to color, not black and white.

• I TRIED TO PRINT SOMETHING AND NOTHING HAPPENED.

Not all printers automatically turn on. Make sure your printer is turned on, then try printing again.

CHAPTER 12 – THE MICROSOFT® MS-DOS® OPERATING SYSTEM AND THE MICROSOFT® MS-DOS® COMMAND PROMPT

This chapter will contain info on the Microsoft® MS-DOS® operating system and some of the different versions of the Microsoft® Windows® operating system.

• WHAT IS THE MICROSOFT® MS-DOS® OPERATING SYSTEM?

Before computers booted up directly to the Microsoft® Windows® XP operating system, they booted up to this software. This is still in use today (in your current version of the Microsoft® Windows® operating system), but it is mostly there for people who have older programs. Most of the time, you only see it when you are not yet in the Microsoft® Windows® XP operating system, such as when you boot from a boot disk.

• WHAT IS A MICROSOFT® MS-DOS® OPERATING SYSTEM COMMAND PROMPT?

A command prompt is that little blinking cursor that you see (when you are in the Microsoft® MS-DOS® operating system area). It's next to the letter of the drive you are on (for example C:/>)

• HOW DO I GET TO A MICROSOFT® MS-DOS® OPERATING SYSTEM PROMPT FROM WITHIN THE MICROSOFT® WINDOWS® XP OPERATING SYSTEM?

In the Microsoft® Windows® XP operating system, you get to a Microsoft® MS-DOS® operating system command prompt by:

1. Clicking "Start"

2. Move your mouse arrow to "Programs".

3. Move your mouse arrow to "Accessories".

4. Move your mouse arrow to "Command Prompt" and click on it.

• TELL ME ABOUT SOME OF THE DIFFERENT VERSIONS OF THE MICROSOFT® WINDOWS® OPERATING SYSTEM?

The most common previous versions of the Microsoft® Windows® operating system were:

The Microsoft® Windows® operating system version 3.1. was one of the first.

Then, in 1995, came the Microsoft® Windows® 95 operating system.

The Microsoft® Windows® 98 operating system came out in 1998.

The Microsoft® Windows® Me operating system came out later.

The Microsoft® Windows® XP operating system came out in 2002.

CHAPTER 13 – TECHNICAL STUFF

Those not used to working with computers might want to be careful with the contents of this chapter.

Let's start with this…

• I NEED A SCREWDRIVER TO USE FOR WORKING ON MY COMPUTER SYSTEM.

Office and computer supply stores sell toolkits for computers, with things specially made to work on computers. It's not a big toolbox. You can carry it around with you.

• I WENT TO A COMPUTER SHOW AND BOUGHT A COMPUTER CASE (TOWER). THERE IS A LOCK ON THE FRONT OF IT. WHAT IS THAT?

It is a keyboard lock. If you put the key (that comes with it) into it and lock it, the keys on your keyboard won't work until you "unlock" it. This is useful if the computer is around many people all day long and you don't want it messed with. Don't lose that key, though.

• I'M GOING TO WORK INSIDE MY COMPUTER. IS THERE ANYTHING I NEED TO DO FIRST?

Yes. Touch the outside of the case (the metal part) to get rid of electricity.

• WHAT IS A PARTITION / PARTITIONING YOUR HARD DRIVE?

A partition is allocating so much space on your physical hard disk for the purpose of splitting it into two (or more) drives. For example, you can take a 40 gigabyte hard drive and split it into a "C:" drive that is 20 gigabytes and a "D:" drive that is 20 gigabytes. You still only have one physical hard drive, but software has now split it up into two drives. Don't worry about it being absolutely necessary to partition your hard drive yourself. Hard drives on new computers are already partitioned by the time you get them. Partitioning can be done on a new hard drive (in which case you want to already have the software; which may very well come with it), or your current hard drive can be re-partitioned as many times as you like.

• WHAT HAPPENS TO THE OTHER HARD DRIVE LETTERS WHEN YOU USE SEVERAL TO PARTITION A HARD DRIVE?

When you partition a hard drive and take up a letter that is already taken (for example – drive "E":), your CD-ROM (or whatever drive "E": was before) becomes drive "F".

• WHEN I PARTITION MY HARD DRIVE, WILL I LOSE THE INFORMATION THAT IS ALREADY ON IT?

Yes, make sure you have backed up your information first before you do this.

• OKAY, YOU SAID THERE WAS SOFTWARE FOR PARTITIONING A HARD DRIVE. WHERE DO I GET THE SOFTWARE FROM?

If it doesn't come with your hard drive, you can download it from the internet. If you download it from the internet, however, make sure that you go to the website of the company that made the hard drive. To find out who made the hard drive, you will have to open up your computer and look on the hard drive label.

• IS THERE ANYTHING ELSE I'M GOING TO NEED WHEN PARTITIONING (OR RE-PARTITIONING) MY HARD DRIVE?

You will need the start-up disk for your operating system and, of course, the partitioning software disk.

• I JUST TRIED TO INSTALL MY NEW CD-ROM DRIVE IN MY COMPUTER, BUT IT STILL ISN'T BEING RECOGNIZED?

Your CD-ROM usually hooks to your motherboard with what they call a ribbon. The ribbon has a red line on one side (which is pin 1) and the other end is pin 40. Make sure that the ribbon (that long, wide whitish cable) is turned around the right way and that the pins on the motherboard, the ribbon and the back of the drive are all oriented correctly.

• HOW MANY PHYSICAL DRIVES CAN I HAVE INSIDE MY COMPUTER?

You can have one floppy drive, up to two hard drives and/or up to two CD-ROM drives or DVD-ROM drives (two drives to a ribbon).

• WHAT IS A BOOT SECTOR?

It is that first little section of your hard drive. As your hard drive gets older, the boot sector will eventually go bad. About the time that this happens, you hard drive will crash and stop being useful to you.

• I HEARD THAT MONITORS PUT OUT A VERY STRONG MAGNETIC FIELD. IS THIS TRUE?

It used to be true, but monitors are much more efficient these days.

AFTERWORD

When computers first started, they took up whole rooms, used punch cards and vacuum tubes. There was no 3-D. No super sophisticated games that had multiple outcomes. They were just a business tool. They weren't the multi-functional items they are becoming today. Computers are indeed a wonderful tool. In today's information oriented society they are a necessity. They have increased in speed, power, versatility and capability rapidly over the past 10 to 15 years. I remember in 1987, when I got my first computer. I had no sound card (all the sound came from the computer speaker) no mouse (that would come a little later; all I had was a joystick), no CD-ROM (The first game I ever had came on two floppy disks).

Today, you can play whole movies, listen to music, restore old photos that you thought were lost forever. There are even ways to help the sight impaired to use a computer.

To some, however, they look very intimidating. Here is this big object. You know it's powerful, you know you can type with it – but how do you get to that point?

I hope that this book will help you with that. This book is for all those people that are a little intimidated by computers, or if you just want an easy reference guide to all those things you keep hearing about in the media – such as hard drives, DVD's, etc. so that you will have the knowledge and confidence necessary to try things you didn't try before.

I've been around computers my whole life. I just wanted to share those experiences with those of you who just needed a little help to get you introduced to the technological world of computers. In other words, take the mystery out of the mysterious. Welcome to the wonderful world of computers.

ABOUT THE AUTHOR

Sean Byerley is a college educated young man who has been around computers for most of his life.

He has over 15 years of computer experience and wanted to pass some of that knowledge on to others who might need it. He has a nice flair to his writing.

He wanted to put out a book about computers and the internet that everyone (even the non computer user) could understand.

www.ingramcontent.com/pod-product-compliance
Lightning Source LLC
Chambersburg PA
CBHW080419060326
40689CB00019B/4300